Graphene Oxide

Wei Gao

Editor

# Graphene Oxide

## Reduction Recipes, Spectroscopy, and Applications

 Springer

*Editor*
Wei Gao
The Department of Textile Engineering Chemistry & Science,
    College of Textiles
North Carolina State University
Los Alamos, NM, USA

ISBN 978-3-319-36463-6          ISBN 978-3-319-15500-5    (eBook)
DOI 10.1007/978-3-319-15500-5

Springer Cham Heidelberg New York Dordrecht London
© Springer International Publishing Switzerland 2015
Softcover reprint of the hardcover 1st edition 2015


Printed on acid-free paper

Springer International Publishing AG Switzerland is part of Springer Science+Business Media
(www.springer.com)

# Contents

# Abbreviations

## Chapter 1

| AFM | Atomic force microscopy |
|---|---|
| CCG | Chemically converted graphene |
| CMG | Chemically modified graphene |
| CNTs | Carbon nanotubes |
| CP/MAS | Cross polarization/Magic angle spinning |
| CVD | Chemical vapor deposition |
| DFT | Density functional theory |
| DRIFT | Diffuse reflectance infrared Fourier transform |
| DSC | Differential scanning calorimetry |
| GO | Graphene oxide |
| PAHs | Polycyclic aromatic hydrocarbons |
| rGO | Reduced graphene oxide |
| SSNMR | Solid state nuclear magnetic resonance |
| STEM | Scanning transmission electron microscopy |
| STM | Scanning tunneling microscopy |
| TEM | Transmission electron microscopy |
| TGA | Thermal gravimetric analysis |
| XPS | X-ray photoelectron spectroscopy |
| XRD | X-ray diffraction |

## Chapter 2

| %A | Percent absorption |
|---|---|
| AOTF | Acousto-optic tunable Filter |
| APD | Avalanche photodiode |
| EMCCD | Electron-multiplied charge-coupled device |
| FFT | Fast Fourier transform |

| FTIR | Fourier transform infrared |
| NIR | Near infrared |
| PL | Photoluminescence |
| QY | Quantum yield |

## Chapter 3

| ATRP | Atom transfer radical polymerization |
| DMF | Dimethylformamide |
| EDA | Ethylenediamine |
| HER | Hydrogen evolution reaction |
| NEXAFS | Near-edge X-ray absorption fine structure |
| NMP | $N$-methylpyrrolidone |
| OER | Oxygen evolution reaction |
| ORR | Oxygen reduction reaction |
| PAA | Poly(allylamine) |
| PEI | Polyetherimide |
| PMMA | Polymethyl methacrylate |
| PVA | Poly(vinyl alcohol) |
| THF | Tetrahydrofuran |

## Chapter 4

| EDLCs | Electric double-layer capacitors |
| GNSs | rGO nanosheets |
| LIBs | Li-ion batteries |
| NPMCs | Nonprecious metal catalysts |
| OER | Oxygen evolution reaction |
| ORR | Oxygen reduction reaction |
| PANI | Polyaniline |
| PEFCs | Polymer electrolyte fuel cells |
| PPy | Polypyrrole |

## Chapter 5

| FO | Forward osmosis |
| ICP | Internal concentration polarization |
| NF | Nanofiltration |
| RO | Reverse osmosis |
| UF | Ultrafiltration |

# Chapter 1
# Synthesis, Structure, and Characterizations

Wei Gao

**Abstract** This chapter introduces a peculiar organic macromolecule previously named as graphitic acid, graphite oxide (GO), or more recently, graphene oxide (GO). It is basically a wrinkled two-dimensional carbon sheet with various oxygenated functional groups on its basal planes and peripheries, with the thickness around 1 nm and lateral dimensions varying between a few nanometers to several microns. It was first prepared by the British chemist B. C. Brodie in 1859 and became very popular in the scientific community during the last decade, simply because it was believed to be an important precursor to graphene (a single atomic layer of graphite, the discovery of which won Andre Geim and Konstantin Novoselov the 2010 Nobel Prize in Physics). Several strategies have been introduced to reduce GO back to graphene; however, in this chapter we will mainly focus on GO itself and, more relevantly, its synthesis, chemical structure, and general characterizations. We emphasize here that, despite its strong relevance to graphene, GO also has its own scientific significance as a basic form of oxidized carbon and technological importance as a platform for all kinds of derivatives and composites that have already demonstrated various interesting applications.

**Keywords** Graphene oxide • Graphite oxide • Hummers • Solid state nuclear magnetic resonance • Magic angle spinning • 2D NMR • Reduced graphene oxide • Polycylic aromatic hydrocarbons

## 1.1 Introduction

Two-dimensional nanomaterials typically refer to flat or slightly corrugated sheets with nanometer thicknesses and infinite lateral dimensions, such as graphene, single-layer boron nitride (BN), molybdenum disulfide ($MoS_2$), etc. The quantum confinement in the thickness direction results in exotic electronic properties and highlighted surface effects that can be useful in sensing, catalysis, energy storage applications, etc. [1].

W. Gao (✉)
The Department of Textile Engineering, Chemistry & Science, College of Textiles,
North Carolina State University, 2401 Research Dr, Campus Box 8301,
Raleigh, NC 27695, USA
e-mail: wgao5@ncsu.edu

© Springer International Publishing Switzerland 2015
W. Gao (ed.), *Graphene Oxide*, DOI 10.1007/978-3-319-15500-5_1

1

They have attracted tremendous attention recently, since they are one of the major categories in nanoscience that were predicted thermodynamically unstable in the free state and has not been well explored. In 2005, Geim's group first reported the experimentally observed room-temperature quantum hall effect on a real piece of graphene, which was obtained by mechanical exfoliation of highly oriented pyrolytic graphite (HOPG) [2, 3]; soon after that, a storm of graphene research dominated the world of carbon nanomaterials science. The term graphene then became a new super star after carbon nanotubes (CNTs) and fullerenes ($C_{60}$) in the carbon world. One of the biggest challenges in graphene research at that time was large-scale production of graphene, since the first method Geim's group adapted was both time consuming and extremely low in yield. Different strategies has been introduced, including metal ion intercalation [4], liquid-phase exfoliation of graphite [1, 5], chemical vapor deposition (CVD) growth [6], vacuum graphitization of silicon carbide (SiC) [7], bottom up organic synthesis of large polycyclic aromatic hydrocarbons (PAHs) [8–10], and of course, chemical reduction of GO [11, 12]. Each strategy has its own advantages and disadvantages; nevertheless, GO was believed to be one of the most promising pathways to graphene, mainly due to its wet chemical processability and large-scale availability to monolayers.

GO is not a naturally occurring compound; the history of GO research can be dated back to over 150 years ago. When it was first made by chemical treatments of graphite with potassium chlorate ($KClO_3$) and fuming nitric acid ($HNO_3$), British chemist Brodie named it graphitic acid or graphite oxide [13], and after graphene research emerged in 2004, people started calling it graphene oxide. Nowadays, a single atomic carbon layer of graphite oxide is considered as graphene oxide, whereas the electronic structures of the two are almost the same due to the weak interlayer coupling [14]. Since most of the experiments on GO were done in wet chemical processes and people are generally dealing with large amounts of GO flakes in solution, it is believed that when GO is dispersed in certain solvents (e.g., water), it is at least partially exfoliated by the solvent molecules and thus can be referred to as graphene oxide. Otherwise, in solid state, GO powder or GO film is basically graphite oxide. In the following content, we will generally abbreviate both graphite oxide and graphene oxide as GO, since there is really no big difference between them, especially in the context of electronic structures and properties (the case is different for graphite and graphene).

In terms of chemistry, GO is a new type of nonstoichiometric macromolecule that is chemically labile and hygroscopic in ambient conditions. However, synthesis of GO has evolved and been modified several times with different chemicals such as potassium permanganate, concentrated sulfuric acid [15], and even phosphoric acid [16]. The resulted compounds differ slightly in their chemical composition depending on the protocol used.

Over the past 150 years, research in GO has been quite limited. Without recent popularity of graphene, researchers would still be confused about its detailed chemical structure. In the past few years, extensive research has been done to elucidate its chemical composition [17, 18], which turns out to be a corrugated carbon sheet with over half of the carbon atoms functionalized with hydroxyl and epoxy groups and edges partially occupied by hydroxyl, carboxyl, ketone, ester, and even lactol structures. Even with those clarifications, the distribution of these groups and the spatial connectivity remain obscure even today. The chemical reduction of GO to graphene is one of the hottest topics in GO research, and the most commonly accepted reagent is hydrazine, as first introduced by Ruoff in UT Austin [12]. Pristine GO is an electrical insulator, and after reduction, it becomes electrically conductive. Several orders of magnitude increase in conductivity are usually observed during reduction processes; however, so far all the graphene materials derived from GO have much poorer crystallinity [19] and carrier mobility than their mechanically cleaved counterparts [2, 3]. Thus, researchers would rather name them reduced GO (rGO), chemically modified graphene (CMG), or chemically converted graphene (CCG). For convenience, we will stick to the term rGO for these materials in the following discussion. The harsh chemical oxidation environment in GO synthesis processes actually creates lots of defects and vacancies within the $sp^2$ carbon lattice, which are almost impossible to recover by subsequent chemical treatments [19]. Taking this into account, GO researchers have switched their interest toward GO and rGO applications. Despite its poor crystallinity, GOs, rGOs, and their derivatives have shown several promising applications in catalysis, composites, energy storage, sensing, water purification, electronics, etc. We note that, in literature, some researchers tend to ignore the differences between rGOs and high-quality graphene and announce their results as graphene properties and applications, while we think it is necessary to put GOs and rGOs into different categories and summarize rGO research into a separated book from those reports on highly crystalline graphene shown elsewhere.

## 1.2  Synthesis

In this section we introduce four different recipes for GO preparation in the chronological order and discuss and compare their chemical processes and product structures in details. The recipes include Brodie method and Staudenmaier method (Sect. 1.2.1), Hummers method and its modification (Sect. 1.2.2), and Tour method and discussions (Sect. 1.2.3).

### *1.2.1  Brodie Method and Staudenmaier Method*

B.C. Brodie, a British chemist in the nineteenth century, prepared the first batch of GO when he was investigating the chemistry of graphite in 1859 [13]. When he added $KClO_3$ into a slurry of graphite in fuming $HNO_3$, he obtained a new batch of compound which later on was determined to contain carbon, oxygen, and hydrogen. He washed the batch free from the salts produced in the reaction, dried it at 100 °C, and again put it under oxidation environment. The appearance of the product changed in the following three repeated treatments and finally resulted in a substance with a "light yellow color" which would not change with any additional oxidation treatment. He emphasized that the product could not be produced by one prolonged treatment, and one had to promote the oxidation process with the restoration of the original conditions each time.

According to his elemental analysis, the molecular formula for the final product was $C_{11}H_4O_5$. Weak acidity and mild dispersibility in basic solution was observed; however, the reflective goniometry characterization failed due to the small size, limited thickness, and imperfect structure. He also reacted the final product with "protochloride of copper and protochloride of tin" to get the GO salts and followed up with detailed analysis of composition and thermal decomposition. Nonetheless, his observations and conclusions were limited by theories and characterization techniques available at that time, leaving a huge space for work and improvement until today.

One of the earliest improvements on Brodie's work was carried out in 1898 by L. Staudenmaier [20, 21]. Two major changes were introduced:

1. Addition of concentrated sulfuric acid to improve the acidity of the mixture
2. Addition of multiple aliquots of potassium chlorate solution into the reaction mixture over the course of reaction

These changes have led to a highly oxidized GO product (composition the same as the final product that Brodie got) in a single reaction vessel, thus largely simplified the process.

However, Staudenmaier method was both time consuming and hazardous: the addition of potassium chlorate typically lasted over 1 week, and the chlorine dioxide evolved needed to be removed by an inert gas, while explosion was a constant hazard. Therefore, further modification or development of this oxidation process was still worth investigation.

## 1.2.2 Hummers Method and Its Modifications

Almost 60 years after the introduction of Staudenmaier's strategy, chemists Hummers and Offeman in Mellon Institution of Industrial Research developed a different recipe for making GO [15]. A water-free mixture of concentrated sulfuric acid, sodium nitrate, and potassium permanganate was prepared and maintained below 45 °C for graphite oxidation. According to their description, the whole oxidation process finished within 2 h, leading to a final product with higher degree of oxidation than Staudenmaier's product (Table 1.1).

However, it was found that Hummers' product usually has an incompletely oxidized graphite core with GO shells, and a preexpansion process is helpful to achieved higher degree of oxidation. First introduced by Kovtyukhova in 1999 [26], a pretreatment of graphite with an 80 °C mixture of concentrated $H_2SO_4$, $K_2S_2O_8$, and $P_2O_5$ for several hours was widely adopted thereafter. The pretreated mixture was diluted, filtered, washed, and dried before the real Hummers oxidation step took place. Later on, it was found that if graphite samples have smaller flake size or have been thermally expanded, the Kovtyukhova pretreatment can be skipped. Other reported modifications also include increase of the amount of potassium permanganate, etc. [11, 27]. Nowadays, this modified Hummers method is the most common recipe used for GO preparation (as described in Fig. 1.1). A typical GO product made by this method consists of thin flakes of GO with 1 nm in thickness

**Table 1.1** Summary of synthetic methods used to prepare GO (Reproduced from ref [25] with permission of the Royal Society of Chemistry)

| Method | Oxidant | Reaction media | Carbon-to-oxygen ratio[a] | Raman spectral $I_D/I_G$ ratio[a] | Charge-transfer resistance $(R_{CT})$[a] (k$\Omega$) | Notes |
|---|---|---|---|---|---|---|
| Staudenmaier [20, 21] | $KClO_3$ | Fuming $HNO_3$ | 1.17 | 0.89 | 1.74 | – |
| Brodie [13] | $KClO_3$ | $HNO_3 + H_2SO_4$ | – | – | – | $KClO_3$ added stepwise rather than in a single bolus |
| Hofmann [22, 23] | $KClO_3$ | Non-fuming $HNO_3$ | 1.15 | 0.87 | 1.68 | – |
| Hummers [15] | $KMnO_4 + NaNO_3$ | Conc. $H_2SO_4$ | 0.84 | 0.87 | 1.98 | Modifications can eliminate the need for $NaNO_3$ |
| Tour [16] | $KMnO_4$ | $H_2SO_4 + H_3PO_4$ | 0.74 | 0.85 | 2.15 | |

[a]Reproduced from a previously reported comparison of the various preparation methods [24]

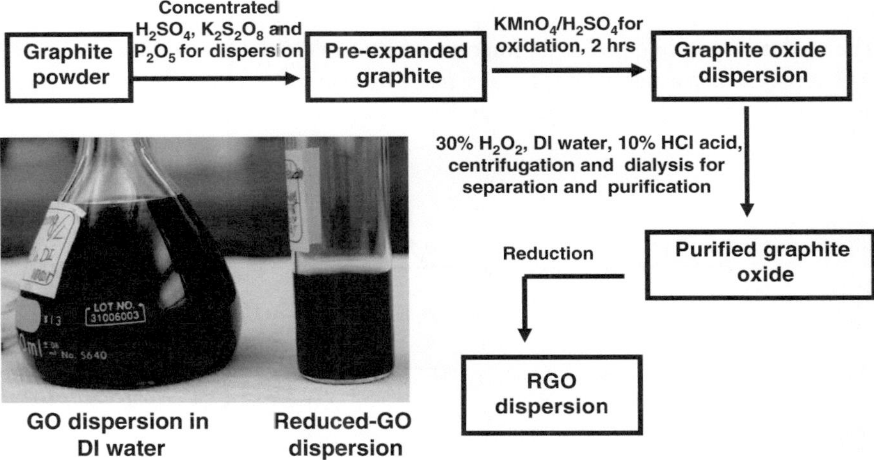


**Fig. 1.1** A schematic of the recipe of the most commonly used modified Hummers method for GO preparation [11]. *Lower left*: photographic images of the final product GO in deionized water (*left*) and the dispersion after hydrazine reduction with ammonia (*right*)

(corresponding to a single layer) and around 1 μm in lateral dimensions on average; meanwhile the chemical composition was determined to be C:O:H = 4:2.95:2.5 [26]. The oxidation degree and yield of GO have been extensively improved when compared with the very first product by Brodie. However, the separation and purification processes in the modified Hummers method are still quite complicated and time consuming.

### *1.2.3  Tour Method and Discussions*

#### 1.2.3.1  Tour Method

As the gold rush of graphene research started in 2004, GO jumped into the center of the carbon nanomaterial research, and lots of publications have emerged talking about its structure, reduction, and applications. In 2010, a new recipe was introduced by Tour's group at Rice University, which avoided the use of sodium nitride and increased the amount of potassium permanganate and also introduced a new acid into the reaction vessel: phosphoric acid [16]. They reported a GO product with higher degree of oxidation made by reacting graphite with six equivalents of $KMnO_4$ in a 9:1 mixture of $H_2SO_4/H_3PO_4$. One of the biggest advantages of this protocol is the absence of $NaNO_3$, thus avoiding generation of toxic gases such as $NO_2$, $N_2O_4$, or $ClO_2$ in the reaction and making it more environmentally friendly. Furthermore, phosphoric acid is believed to offer more intact graphitic basal planes, and the final yield is much higher than that in Hummers method. A comparison among these protocols is shown in Fig. 1.2.

#### 1.2.3.2  Discussions

The source of graphite is also an important factor in GO fabrication [28, 29]. Chen et al. reported disparities in GO products made from five different commercial sources of graphite following a modified Staudenmaier protocol. These graphite samples,

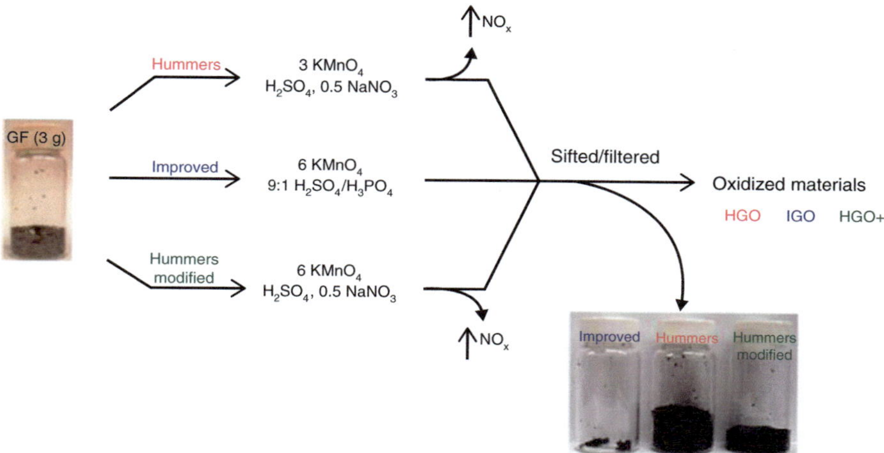


**Fig. 1.2** A comparison of procedures and yields among different GO preparation recipes (Reprinted with permission from ref. [16]. Copyright (2010) American Chemical Society)

though sharing the same idealized chemical structure, have drastic differences in their grain sizes, dispersibilities, reactivities, and especially the propensity for oxidation. As a result, the defect densities in crystalline structures, which could serve as starting sites for chemical oxidation, differ a lot in these precursors. In addition, due to the inherent defects and complexity of the structure, the precise oxidation mechanism in those reactions is hard to elucidate. Chen et al. observed that the product from the highest crystallinity graphite offers best transport for electrons and holes [28]. Kim et al. also prepared GO from three different graphite sources with modified Hummer's method and observed disparities in sheet size distributions [30]. Interestingly, graphite nanofibers have also been used as starting material to make GO, and resulted GO nanosheets are more uniform in size distribution. The coin-stacked graphene planes along the length of the fibers are believed to play an important role here, and tunability of the GO size upon oxidation time was also observed [31].

So far, two different combinations of oxidation reagents have been used to oxidize graphite to GO, including potassium chlorate with nitric acid and potassium permanganate with sulfuric acid (both acids are in the most concentrated state). In literature, nitric acid has been reported to react with aromatic carbon surfaces such as carbon nanotubes [32] and fullerenes [33], resulting in various oxygenated functional groups such as carboxyls, lactones, and ketones and, meanwhile, releasing of toxic gases like $NO_2$ and $N_2O_4$. Similarly, potassium chlorate provides its oxidation capability by in situ generating dioxygen that is very reactive [34]. When Brodie method and Staudenmaier method were introduced for GO synthesis, these chemicals were believed to be the strongest oxidizers available. As for the second combination ($KMnO_4$ and $H_2SO_4$), permanganate ion is also a typical oxidation reagent. The reactivity of $MnO_4^-$ can only be activated in acidic media, mainly described by the formation of dimanganese heptoxide from $KMnO_4$ in the presence of strong acid as given in Scheme 1.1 [35]:

The transformation of $MnO_4^-$ into a more reactive form $Mn_2O_7$ will certainly help oxidize graphite, but the bimetallic form of manganese oxide has been reported to be detonate when heated up to 55 °C or when reacted with organic compounds [35, 36].

The final acid that has been introduced into the GO synthesis is phosphoric acid, which is also believed to have an advantage of offering more intact $sp^2$ carbon domains in the basal planes of the final products [37]. Figure 1.3 shows a possible explanation for this advantage, as adapted from ref. [37]. The formation of the five-membered phosphor ring helps prevent the further oxidation of the diols.

$$KMnO_4 + 3H_2SO_4 \rightarrow K^+ + MnO_3^+ + H_3O^+ + 3HSO_4^-$$
$$MnO_3^+ + MnO_4^- \rightarrow Mn_2O_7$$

**Scheme 1.1** Formation of dimanganese heptoxide from $KMnO_4$ in the presence of strong acid (Reprinted with permission from ref. [35]. Copyright (1982) American Chemical Society)

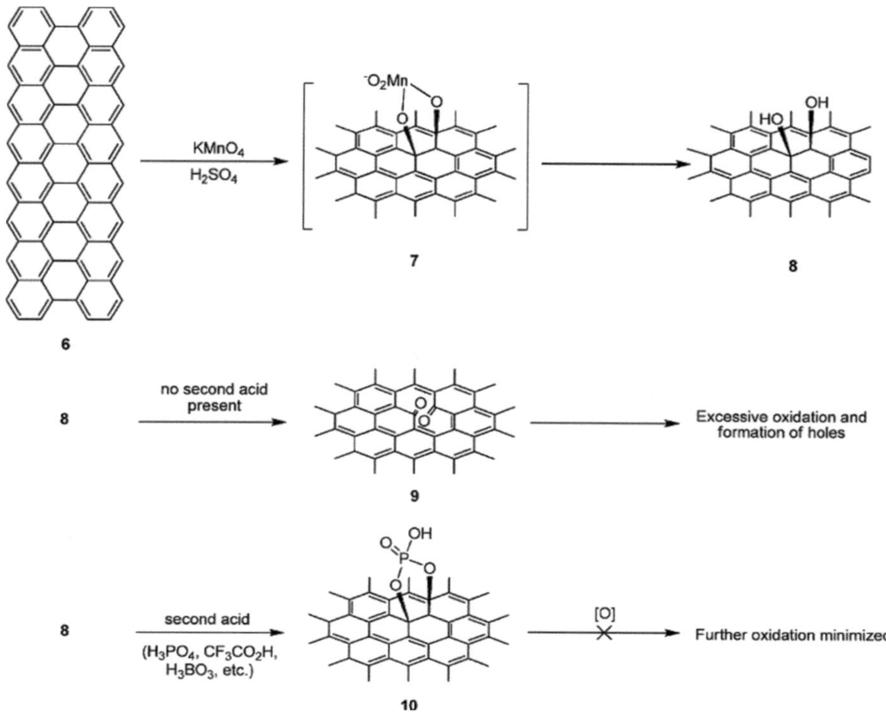


**Fig. 1.3** Proposed mechanisms for the effect of the second acid in prevention of over-oxidation of the $sp^2$ carbon network once they have formed the vicinal diols (Reprinted with permission from ref. [37]. Copyright (2010) American Chemical Society)

In addition, they also claimed that in freshly made GO, organosulfate moieties exist as one of the major functionalities [38].

Purification is another important but tedious step in GO fabrication, since all of these protocols require long washing, filtration, centrifugation, and dialysis step. It has been reported that GO contaminated with potassium salts is highly flammable, which poses a fire hazard. The volume expansion and gelation observed during water washing of GO significantly slow down the purification process, and substitution with HCl acid and acetone was introduced by Kim et al. [39].

In summary, at least four different recipes have been introduced for GO preparation in the history of GO synthesis, and improvements in oxidation, simplicity, yield, and the product qualities have been demonstrated. Today, making a batch of GO is no longer a problem, which thus has facilitated the rush of GO research; however, we are still lack of the basic understanding of the oxidation processes and detailed mechanisms, which prevents chemically engineering and manipulation of a variety of reactions to tackle critical technology issues, such as bandgap tuning, size distribution control, edge structure selection, etc.

## 1.3  Spectroscopic Characterizations and Chemical Structure

As a unique form of oxidized carbon, GO lies beyond the scope of organic compounds and large polycyclic aromatic hydrocarbons (PAHs), thus making it quite interesting and challenging to characterize this peculiar structure. In this section, we will start with spectroscopic characterization results on GO and then try to clarify its molecular structure.

### *1.3.1  Spectroscopic Characterizations*

#### 1.3.1.1  Solid-State $^{13}$C Nuclear Magnetic Resonance (SSNMR) Spectroscopy

GO sheets are gigantic molecules that fall into the category of colloids. GO dispersions in water are much too diluted to be analyzed by liquid-phase NMR. In literature, the most powerful and precise technique to characterize GO is solid-state $^{13}$C NMR [40]. Due to the low natural abundance of $^{13}$C (1.1 %), the signal to noise ratio in the measurement of regular samples is quite low, and long acquisition time is usually required for good quality data. Therefore, in 2008, Ruoff and coworkers prepared a $^{13}$C-enriched GO sample and clarified its chemical structure to a new level [17]. According to their analysis, cross polarization/magic angle spinning (CP/MAS) experiments displayed three broad resonances at 60, 70, and 130 ppm in the $^{13}$C NMR spectrum of GO. Figure 1.4 shows typical one-dimensional (1D) and two-dimensional (2D) $^{13}$C pulse spectra from a $^{13}$C-labeled sample (adapted from ref. [17]), clearly demonstrating both the chemical structure assignments of major peaks and the spatial proximity between the $sp^2$ carbon, epoxy carbon, and hydroxyl carbon atoms (as indicated by the green, blue, and red circles in Fig. 1.4b).

Interestingly, although the isotopical labeling of GO greatly improved the SSNMR analysis resolution, there still remained some unassigned peaks in this work. Later on, further reports came out with more detailed assignments of those peaks, such as a new identification of the 101 ppm peak, which has long been ignored by previous researchers. The 101 ppm peak probably comes from the five- or six-membered-ring lactol structure as shown in Scheme 1.2. Figure 1.5 shows a typical comparison of a $^1$H–$^{13}$C cross-polarization spectrum and a direct $^{13}$C pulse spectrum obtained from unlabeled GO, with quantitative data on the relative ratio of all these functionalities to be 115 (hydroxyl and epoxy):3 (lactol O–C–O):63 (graphitic $sp^2$ carbon):10 (lactol + ester + acid carbonyl):9 (ketone carbonyl) [18].

More advanced SSNMR techniques were later used for GO characterization, including 2D $^{13}$C double-quantum/single-quantum (2Q/SQ) correlation SSNMR, 2D $^{13}$C chemical shift anisotropy (CSA)/isotropic shift correlation SSNMR, and 2D triple-quantum/single-quantum (3Q/SQ) correlation SSNMR [41]. The 2Q/SQ spectrum eliminates diagonal signals in 2D spectrum, offering a clearer correlation signal between $^{13}$C–OH carbon and $^{13}$C–O–$^{13}$C carbon (Fig. 1.6). The 3Q/SQ spectrum

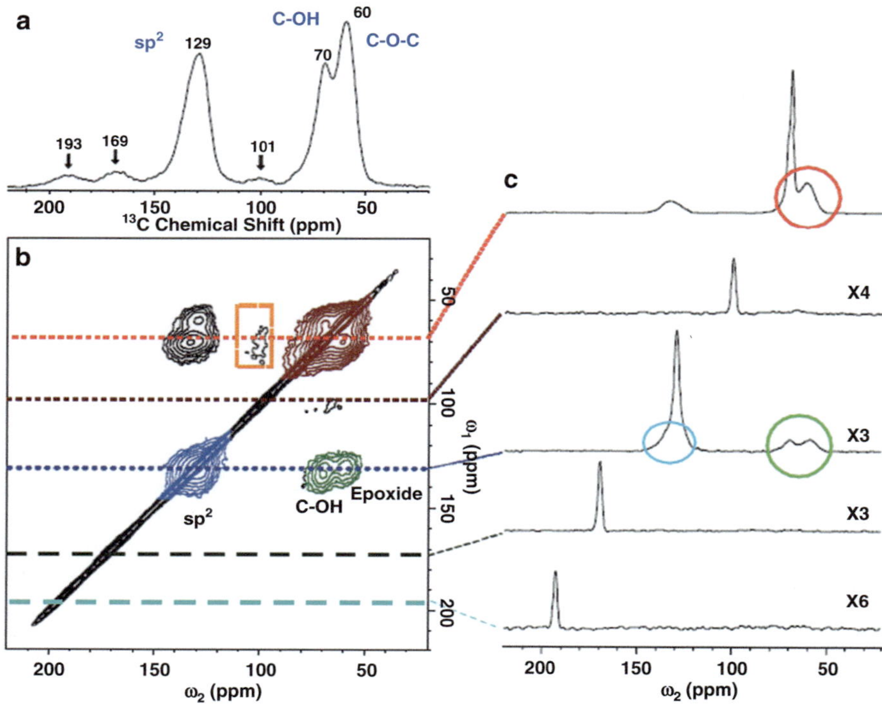



**Fig. 1.4** (**a**) 1D $^{13}$C MAS and (**b**) 2D $^{13}$C/$^{13}$C chemical-shift correlation solid-state NMR spectra of $^{13}$C-labeled graphite oxide with (**c**) slices selected from the 2D spectrum at the indicated positions (70, 101, 130, 169, and 193 ppm) in the ω1 dimension. All the spectra were obtained at a $^{13}$C NMR frequency of 100.643 MHz with 90 kHz $^{1}$H decoupling and 20 kHz MAS for 12 mg of the sample. In (**a**), the $^{13}$C MAS spectrum was obtained with direct $^{13}$C excitation by a $\pi/2$-pulse. The recycle delay was 180 s, and the experimental time was 96 min for 32 scans. In (**b**), the 2D spectrum was obtained with cross polarization and fpRFDR $^{13}$C–$^{13}$C dipolar recoupling sequence. The experimental time is 12.9 h with recycle delays of 1.5 s and 64 scans for each real or imaginary t1 point. A Gaussian broadening of 10 Hz was applied. The *green*, *red*, and *blue* areas in (**b**) and *circles* in (**c**) represent cross peaks between $sp^2$ and C–OH/epoxide (*green*), those between C–OH and epoxide (*red*), and those within $sp^2$ groups (*blue*), respectively (From ref. [17]. Reprinted with permission from AAAS)

offers coherence correlation between three different carbons, thus providing a large amount of information regarding the connectivity between differently functionalized carbon atoms as well as the distribution of those functional groups on GO surfaces. Interestingly, theoretical simulation was used simultaneously to fit these data with simplified GO structure models, and satisfying fitting was obtained by ab initio calculations with the structure model A (with only 1,2-ether) shown in Fig. 1.6d.

All these analysis significantly helped identify GO's chemical structure, leading to improved clarity in its chemical compositions. However, it is also noteworthy to point out here that all those SSNMR characterizations were done on GO samples made from the modified Hummers method and GO products from other methods do differ in the relative ratio of these functionalities [16].

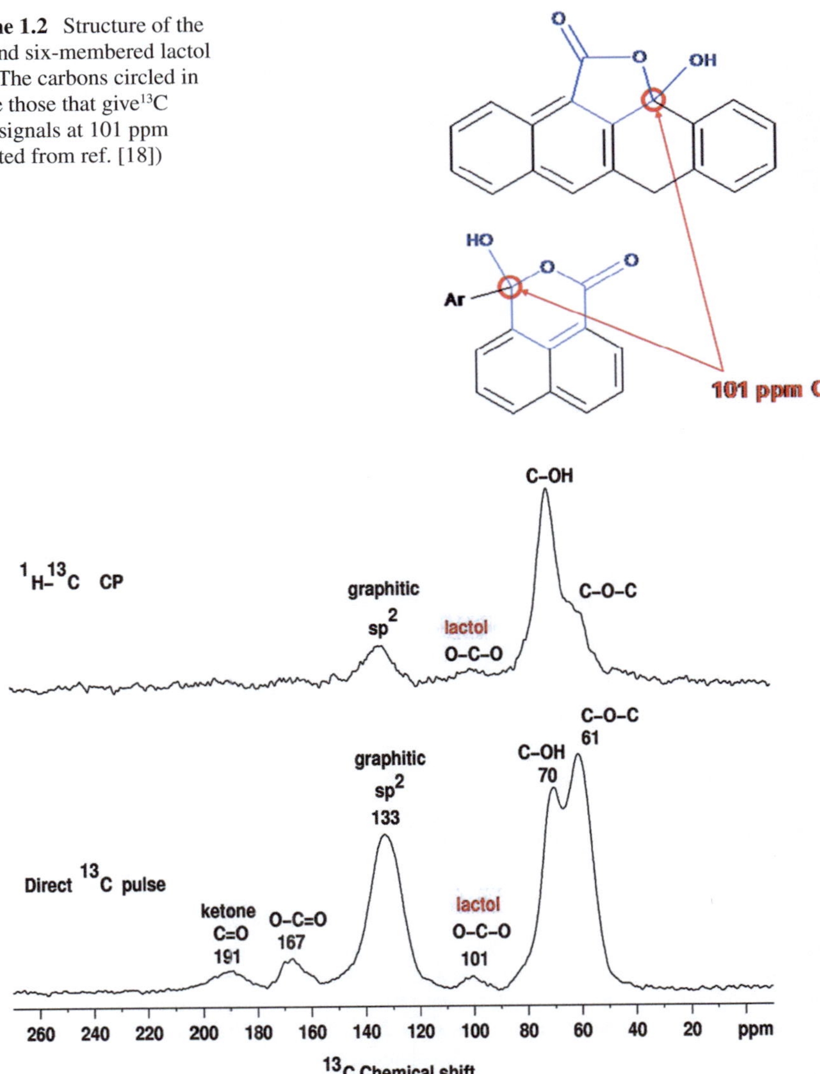



**Fig. 1.5** A $^1$H–$^{13}$C cross polarization (CP) spectrum of GO obtained with 7.6 kHz MAS and a contact time of 1 ms (67,000 scans, *top*) and a direct $^{13}$C pulse spectrum obtained with 12 kHz MAS and a 90° $^{13}$C pulse (10,000 scans). The peak at 101 ppm is caused by the carbons of five- and six-membered-ring lactols (adapted from ref. [18])

### 1.3.1.2  Diffuse Reflectance Infrared Fourier Transform (DRIFT) Spectroscopy

Besides SSNMR, DRIFT spectroscopy is another powerful tool to detect chemical functionalities in oxidized carbons [42, 43]. For samples of GO, since it is hygroscopic, the large amounts of water molecules (up to 25 wt%) adsorbed in the

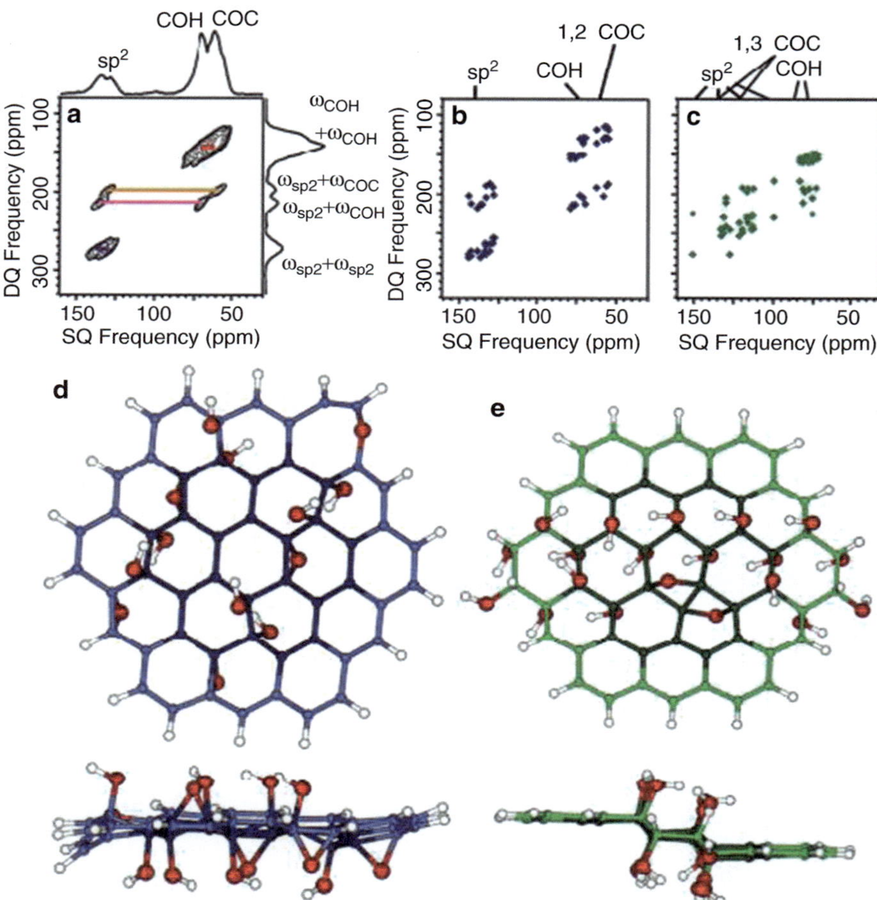

**Fig. 1.6** (**a**) Experimental 2D $^{13}$C DQ/SQ correlation SSNMR spectrum of uniformly $^{13}$C-labeled GO using $^{13}$C–$^{13}$C *J* coherence transfer. Fast recycling with short recycle delays of 0.3 s and low-power (7 kHz) decoupling was used. Signal assignments in (**a**) are those made in ref [10] and confirmed here. The carrier frequency was set at 211.17 ppm. (**b, c**) Predicted DQ/SQ correlation spectra based on isotropic chemical shifts calculated for (**b**) model A and (**c**) model B. (**d, e**) Structural models for (**d**) model A and (**e**) model B from the top and (*bottom*) side views. Carbons in (**d, e**) are color coded to match the spectra in (**b, c**). *Red* and *white spheres* denote O and H, respectively. $^{13}$C at the edge of the models (*light blue* or *green*) was not included in (**b, c**). The calculated spectrum (**b**) based on model A well reproduced the experimental spectrum (**a**) (Adapted from ref. [41] with permission. Copyright (2010) American Chemical Society)

structure have become the biggest obstacle to FTIR analysis. Therefore, Décány and coworkers reported a detailed DRIFT analysis of deuterated GO samples prepared with the Brodie method, and they were able to assign most of the peaks observed in the spectra and distinguish the signals of hydroxyl groups from that of others, thus providing a strong clarification of all the previous debated assignments

**Table 1.2** IR peak positions (in cm) of air-dry GO-1 (GO-1/H$_2$O), GO-4 (GO-4/H$_2$O), and deuterated GO-1 (GO-1/D$_2$O) and their assignments[a] (Reprinted with permission from ref [43]. Copyright (2006) American Chemical Society)

| | GO-4/H$_2$O | GO-1/D$_2$O | Assignment |
|---|---|---|---|
| | | 3,210 | $\nu_{OH}$ in HDO |
| | | m, br | |
| 3,630* | 3,634* | 2,680* | $\nu_{OH}$ in C–OH/$\nu_{OD}$ in C–OD |
| m, sp | w, sh | m, sh | |
| 3,490* | 3,490* | 2568* | $\nu_{OH}$ in C–OH/$\nu_{OD}$ in C–OD |
| s, sp | m, sp | vs,br | |
| 3,210* | 3,190* | 2,396* | $\nu_{OH}$ in H$_2$O/$\nu_{OD}$ in D$_2$O |
| s, br | s, br | vs, br | |
| 2,814 | 2,780 | | $\nu_{OH}$ in dimeric COOH |
| w, br | w, sh | | |
| 1,714 | 1,738 | 1,716 | $\nu_{C=O}$ |
| m, sp | vs, sp | m, sp | |
| 1,616* | 1,616* | 1,196* | $\beta_{OH}$ in H$_2$O/$\beta_{DO}$ in D$_2$O |
| m, sp | s, sp | m, sp | |
| 1574 | | 1,574 | Aromatic vC=C |
| w, sh | | m, sp | |
| 1,368* | 1,370* | 968* | $\beta_{HO}$ in C–OH/$\beta_{OD}$ in C–OD |
| s, br | s, br | s, br | |
| | | 1,384 | Organic carbonate |
| | | m, sp | |
| | 1,196 | | Phenolic OH |
| 1,064 | 1,066 | 1,064 | Skeletal modes of $\nu_{C–C}$ and $\nu_{C–O}$ bonds (*m, sp*) |
| 968 | 978 | 968 | |
| 828 | 828 | 828 | |
| 698 | 696 | 698 | |

Band intensities and widths are classified as w (weak), m (medium), s (strong), vs (very strong), sh (shoulder), sp (sharp), and br (broad)

[a]*Asterisks* (in the same row) designate isotopomer peak pairs

in GO FTIR spectrum (Table 1.2) [42, 43]. Deuterium exchange over GO imposes red shifts of all hydroxyl-related bands; thus, one can easily distinguish between the C–OH and COOH stretching vibrations, and a new band was also uncovered around 1,384 cm, which was hidden in the spectrum of air-dry GO and was attributed to organic carbonates. The possible functional moieties in GO includes carboxyl, lactone, phenol, lactol, chromene, ketone, etheric rings, and organic carbonate, but definitely no pyrones [42, 43]. Shoulder at 1,574 cm in less oxidized GO (GO-1) was possibly assigned to aromatic region vibration (Fig. 1.7a). Broad peak at 1620 cm has been assigned to $\beta_{HOH}$ bending in water (Fig. 1.7b, c), since the signals were depressed in anhydrous GO (Fig. 1.7b) and almost disappeared in deuterated GO DRIFT spectra (Fig. 1.7c).

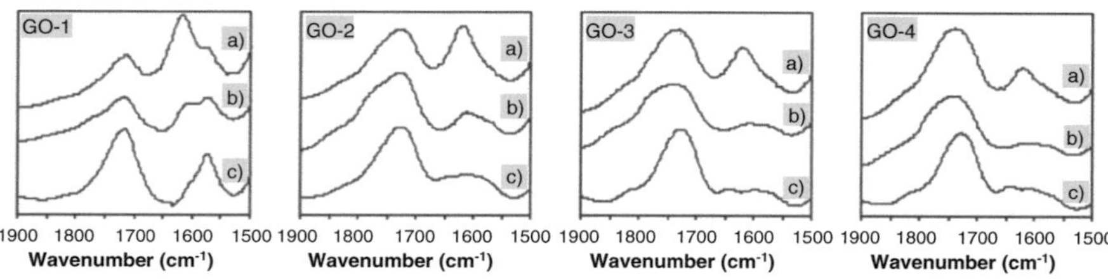



**Fig. 1.7** 1,900–1,500 cm region of (**a**) air-dry GO, (**b**) anhydrous GO, and (**c**) deuterated GO DRIFT spectra. The ordinate scale is arbitrary (Reprinted with permission from ref. [43]. Copyright (2006) American Chemical Society)

### 1.3.1.3   Raman, UV–Vis, XRD, and XPS

As a pseudo-ordered structure, the surface of GO is covered nonuniformly with those oxygenated groups, leaving approximately 2–3 nm $sp^2$ carbon clusters isolated within the $sp^3$ carbon matrix, as can be easily verified by Raman spectroscopy [19, 44], scanning tunneling microscopy [45, 46], high-resolution transmission electron microscopy [47, 48], and transport studies [49, 50].

The Raman spectrum of GO basically consists of a broad D peak ($\approx$1,350 cm) and G peak ($\approx$1,580 cm), a wide weak bump extended from 2,681 to 3,050 cm. The D/G ratio is around 0.95, indicating a large amount of defects within the crystal lattice. However, upon chemical reduction, an increase in D/G ratio is usually observed, the explanation for which remains ambiguous.

UV–Vis spectrum of GO in water has two featured peaks around 233 nm due to $\pi$–$\pi$* transition of C=C bonds and broad shoulder between 290 and 300 nm assigned as $n$-to-$\pi$* transition of C=O bonds [51]. Upon reduction, a red shift of the first peak and the disappearance of the second one were usually observed.

X-ray diffraction (XRD) analysis of GO powder shows a prominent but somewhat broad peak round 11°, the position of which can be easily influenced by oxidation and hydration degree of the GO sample and, of course, the environment humidity during the measurement. The reported interlayer distance of GO samples varies from 5.97 [42] to 9.5 Å [16].

X-ray photoelectron spectroscopy (XPS) analysis of GO powder also offers two broadened and overlapped peaks centered around 284 and 286 eV, corresponding to $sp^2$ carbon and oxidized carbon, respectively. Some studies try to deconvolute these peaks into different oxidation functionalities, while we think XPS is at most a semi-quantitative analysis technique, and this kind of analysis is going beyond the resolution limit of the instrument. In addition to that, to accurately assign to peak position, it is necessary to have stable elements (such as gold) as reference. Many researchers are used to take carbon signal as the major reference in XPS, which is definitely questionable in XPS analysis of GO and rGOs.

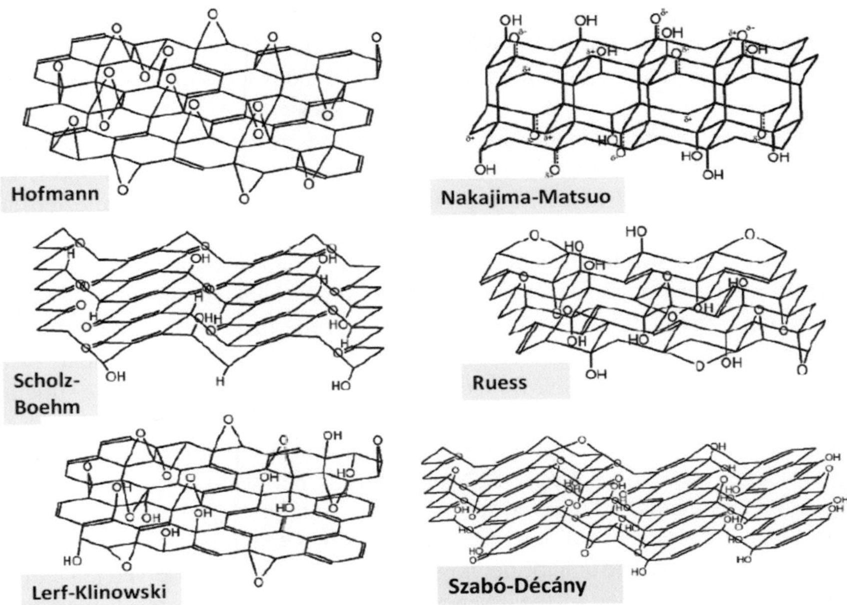



**Fig. 1.8** Proposed structure models for GO (Reprinted with permission from ref. [43]. Copyright (2006) American Chemical Society)

## 1.3.2 Chemical Structure

Based on the characterizations listed above (mainly SSNMR and DRIFT analysis data), at least six different structure models for GO have been suggested and the precise chemical structure of GO remains controversial (Fig. 1.8) [43].

The earliest model by Hofmann and Holst presented in 1939 consists of epoxy groups spreading across the basal planes of graphene, with C/O ratio of two [23]. The model was modified by Ruess in 1946 with the introduction of hydroxyl groups into the lattice and also the corrugating of the basal plane [52]. Different from the Hofmann model, the Ruess model prefers 1,3-ether on a cyclohexane ring with the four-position hydroxylated and also stoichiometrically regular. Ruess's suggestion was supported by the observed structure of poly(carbon monofluoride) $(CF)n$ [53] later by Mermoux in 1991. The existence of hydroxyl groups in this model accounts for the hydrogen content in GO for the first time; in 1957, Clauss and Boehm supplemented this contribution with C=C bonds, ketone, and enolic groups, as well as the carboxylic groups on the edges [54]. More than a decade later, Scholz and Boehm reconsidered the stereochemistry of this model and modified it into a corrugated carbon layers consisting of alternately linked ribbons of quinoidal structure and opened cyclohexane rings in chair conformation [55]. They completely removed epoxy and ether structure from this model and put hydroxyl groups at the four-position of 1,2-oxidized cyclohexane rings. On the other hand, the Ruess model was still a possibility; in 1988, Nakajima and Matsuo proposed a stage 2 type model $(C_2F)n$ in graphite fluorinated product and tried to make the oxide analog for GO [56].

Yet most of the above models have been supplanted by the two most recent models named after Lerf and Klinowski [40, 57, 58] and Szabó and Décány [43], respectively. In the Lerf–Klinowski model, the periodicity in the structure was rejected and substituted with a nonstoichiometric amorphous alternative. In their studies, SSNMR technique was for the first time introduced for GO structural characterization, while all the previous reports were simply based on elemental analysis, reactivity observations, and XRD data. Obviously this is an important milestone in the GO structure debate, and as discussed in the previous section, the assignments of all those CP/MAS signals in SSNMR spectra soon helped to clarify the basic structure information for GO.

Lerf and coworkers have presented many detailed discussions and analyses of GO structure, based on not only SSNMR analysis but also GO reactivity with a variety of compounds and infrared spectroscopic data. First of all, they tried to have a Diels–Alder-type cycloaddition reactions $(4+2)$ on GO (conjugated double bonds should react) with maleic anhydride; however, no reaction was observed [57]. In the context of the distribution of the alkenes on GO basal plane, the result they got is definitely inconclusive, since the lack of reactivity of GO with dienophile could be due to the complexity on the GO local environment such as the steric effect of epoxy and hydroxyl groups. Later, they suggested that alkenes (C=C) in GO are probably either aromatic or conjugated, based on the logic of isolated double bonds could not survive the harsh oxidation environment applied when making GO [40]. Secondly, in combine with the interpretation on FTIR data earlier, they proposed that keto groups are more favored at the periphery of GO than carboxylic acids [59]. The acidity in GO was further explained by them with a keto–enol tautomerization and the proton exchange on the enol site. The keto form is supposed to be thermodynamically more favored; however, $\alpha,\beta$-unsaturated ketones that are present in aromatic regions would favor phenoxide product and allow proton exchange (also known as phenol–quinone exchange, shown in Scheme 1.3). Thirdly, they observed strong hydrogen bonding between GO flakes themselves and water molecules, as indicated by the constant full width at half maximum height of the water peak in the $^1H$ NMR spectrum [58].

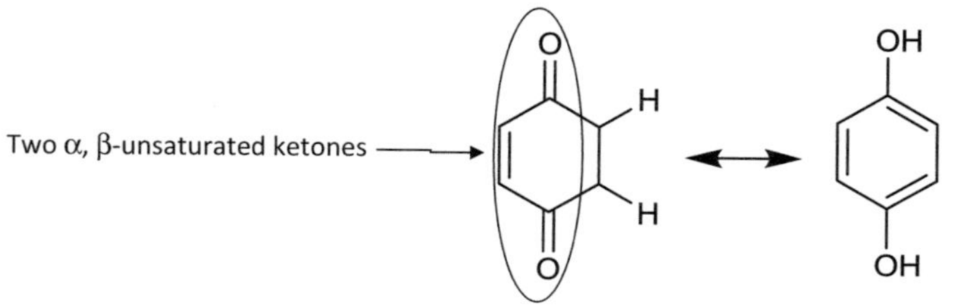

Two α, β-unsaturated ketones

**Scheme 1.3** Schematics of phenol–quinone exchange, favoring the forward direction due to the aromaticity in phenol structure

The Szabó–Décány model was another well-recognized structure for GO, which adopted the logic in Ruess and Scholz–Boehm models and stuck to the corrugating nature of the carbon network. It followed the basic framework in Scholz–Boehm model, while adding 1,3-ethers into the structure and extending the trans-linked cyclohexyl networks [43]. In 2008, Cai et al. prepared $^{13}$C-labeled GO sample and conducted 1D and 2D SSNMR analysis on it. According to their conclusion, only the Lerf–Klinowski model and the Szabó–Décány model are possible [17].

Later on, ab initio chemical shift calculations were used to simulate the SSNMR signals in GO by Ruoff's group [41], and an experimental 2D $^{13}$C double-quantum/single-quantum correlation SSNMR spectrum of $^{13}$C-labeled GO was compared with spectra simulated for different structural models using ab initio geometry optimization and chemical shift calculations. According to their conclusion, only the Lerf–Klinowski model fits best with the experimental data; furthermore, all the previous proposed models were excluded. This is definitely another big step forward in the clarification of the GO structure; however, we need to point out that, due to the size limitation in the theoretical modeling, the Lerf–Klinowski model used to simulate the NMR spectra in this paper is quite simplified, and thus, trivial details of the structure on the edges as well as the precise distributions of those functional groups are still unclear.

In addition to the existing models, Dimiev et al. suggested a new "dynamic structure model" for GO in 2012 [60]. In their explanations, the acidity of GO originated from its interaction with water, and prolonged exposure of GO to water gradually degrades GO and converts it into humid acid-like structures. This claim indicated that GO dispersion would degrade over time, which is in fair agreement with the metastability of GO claimed by Kim et al. later in the same year, although the degradation mechanism in the latter case was suggested totally different (see Sect. 1.3.3 for details) [61]. Interestingly, the purified GO dispersion in DI water was reported to be a nematic liquid crystal [30, 62]. Kim et al. showed that its liquid crystallinity could be maintained with addition of nanoparticles or polymers. The orientation could also be manipulated by external magnetic field or mechanical deformation [30].

In 2011, Rourke et al. first pointed out the existence of oxidation debris (OD) on GO surface, which is around 1 % by mass in GO, amorphous and rich in acidic groups, can be removed from as-prepared GO with OD desorption in 0.1 M NaOH solution [63–65]. The purified GO (after OD removal) exhibited dramatic difference in properties, such as solubility in water, UV, and FTIR absorption. This new result indicates that GO, as viewed earlier as a pseudo-ordered macrostructure, might be a composite of OD and less oxidized graphene sheets. Since this statement is in fair agreement of the observation we had in high-resolution STEM (see Sect. 1.4.2) as well as other reports [66], we consider this new statement regarding GO structure is worth of further investigation.

Another important statement developed recently is the presence of sulfur functionalities in GO. Dimiev et al. first demonstrated the presence of covalent sulfates on GO platform in 2012 [38]. The same group of authors later on quantitatively determined the contribution of sulfates to integral GO acidity [60]. In 2013, Eigler et al. also reported sulfate species identified as covalently bound to GO and still present after extensive aqueous workup [67]. In addition, they were able to exclude the sulfonic groups in GO after aqueous workup. Of course, sulfur-containing moieties can only exist in GO products made with sulfur-containing precursors, namely, Hummers method and Tour method, both of which involve sulfuric acid. According to the estimation of Eigler et al., for each sulfate group there are 20 carbon atoms, which explains the acidity of GO in water (p$K$a 3–4). These organosulfate groups are much more susceptible to hydrolysis as compared to their carboxylate analogs, which were conventionally considered as the major source of acidity in GO.

Unfortunately, to date, the precise structure of GO remains elusive. Major reasons include sample-to-sample variability due to different synthesis methods and degrees of oxidation; amorphous, nonstoichiometric nature of GO; and limited resolution in the major characterization techniques such as SSNMR and FTIR. In this case, the term "graphite oxide" may refer to a family of different compounds with certain discrepancies in the functional group distributions and relative content.

### 1.3.3  Structure Degradation

In 2012, Riedo et al. reported the room-temperature metastability of graphene oxide films, claiming that no matter what the detailed structure is for GO, it is constantly degrading in the ambient environment until it reaches a quasi-equilibrium state, with a characteristic relaxation time of about 1 month [61]. The degraded GO loses epoxides and gains hydroxyl groups, while the whole degradation process is limited by the availability of hydrogen in GO samples [61]. The process of the degradation is simulated as shown in Fig. 1.9.

This report has sparked enormous attention among GO research community, and various subsequent investigations came out within the past 2 years [68–72].

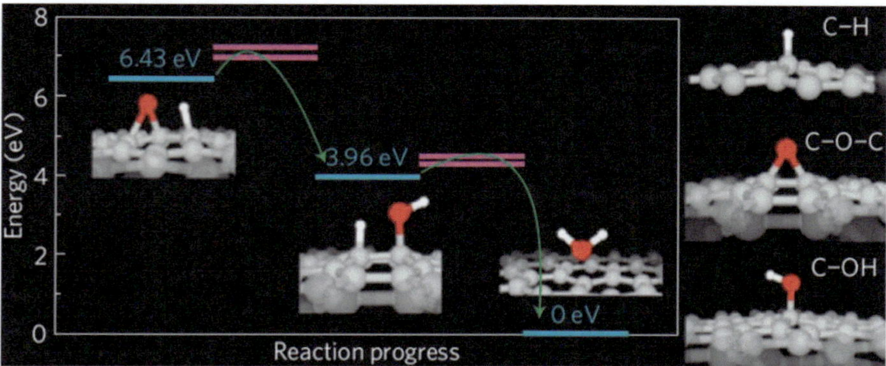


**Fig. 1.9** Density functional theory (DFT) energy diagram for GO reduction (Reprinted by permission from Macmillan Publishers Ltd: Nature Materials [61], copyright (2012))

One interesting example is shown by Chua and Pumera [72], where they differentiate the stability of graphite oxide and graphene oxide under influences of light and atmosphere. In their conclusions, the quasi-equilibrium states of both can be maximized in O/C ratios in dark for graphite oxide and in inert atmosphere for graphene oxide. This work gave helpful insights on how GO researchers should store and preserve their GO samples to maintain the maximum level of oxidation. Another interesting protocol to overcome this degradation is by ozone treatment [71]. Gao et al. reported the increased level of oxidation on degraded GO with bubbling of fresh $O_3$ into the GO dispersion in deionized (DI) water. In addition to the enhanced (or recovered) oxidation, higher pinhole density and smaller flake size were also observed by the authors.

## 1.4    Morphological Characterizations and Thermal Stability

Typically, transmission electron microscopy (TEM) and scanning transmission electron microscopy (STEM) are used to characterize the microscopic nature of GO, especially the localized environment on its basal planes and edges. Atomic force microscopy (AFM) is commonly used in many literatures to characterize the thickness of the GO flakes obtained by authors, in order to justify the single-layer or few-layer statement. However, scanning tunneling microscopy (STM) are rarely used on GO, mainly due to the lack of electrical conductivity, as well as limited access to instrumentation in some cases. Thermal stability of GO is thoroughly studied in literature, since it is also correlated with GO's transformation back to graphene, which is the initial motivation for GO research a decade ago. Reports focusing on heating GO to achieve its conversion to graphene are summarized in Chap. 3 (Table 3.1).

### *1.4.1    Transmission Electron Microscopy (TEM) and Scanning Transmission Electron Microscopy (STEM)*

Given the obscure nature of GO structure, microscopic characterizations of this group of two-dimensional materials sparked high attention in literature [48, 73]. High-resolution TEM or STEM imaging can greatly help us understand the atomic structure and electronic properties of GO, although discrepancies have been observed among different reports. For example, stable images of GO were collected and shown under 80 kV electron beams [48], whereas in some cases only a maximum of 60 kV can be used for GO imaging [71]. In our experiments, we noticed GO is not stable under high-voltage beam (>60 eV). The oxidized moieties tend to move and deposit at where the electron beam is focused at. Here we show a high-resolution STEM images of GO obtained at 60 eV by David Cullen and Karren

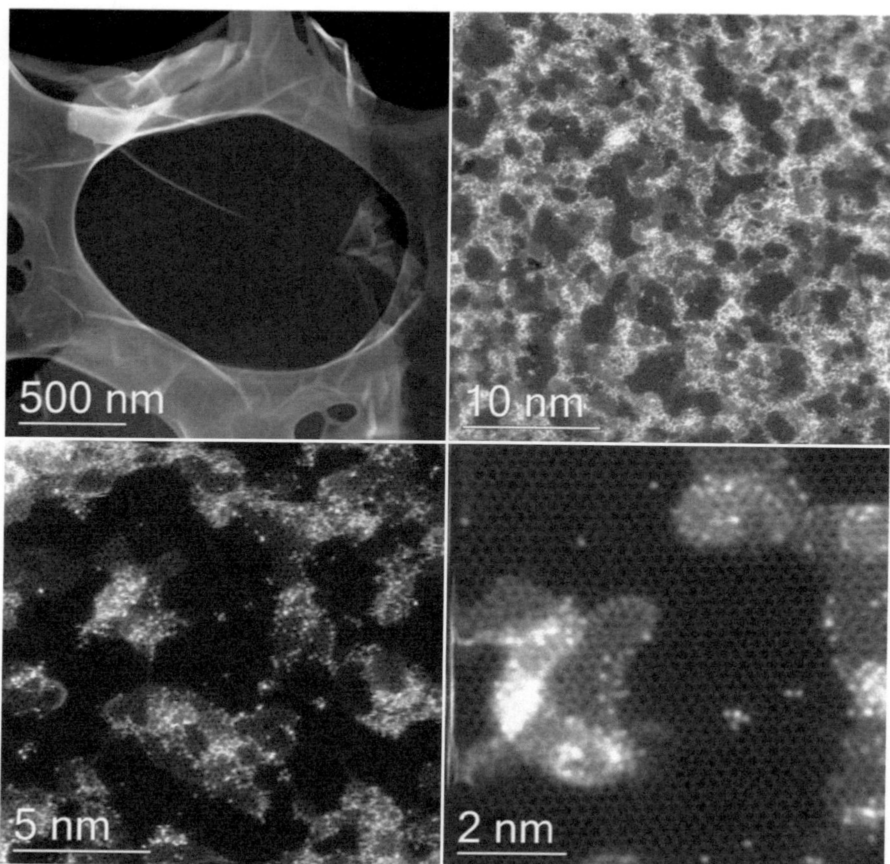


**Fig. 1.10** Annular dark field (ADF)–STEM images of GO samples made with modified Hummers method. The GO basal planes here were largely covered with oxidation moieties, whereas the $sp^2$ carbon backbones were also pretty clearly observed. In addition to the graphitic and oxidized carbons, pinholes were also present in the sample. The individual bright spots in the ADF images were identified to be silicon, a contaminant that is commonly found in graphene and graphene oxide. Because of their much larger atomic number, the Si atoms dominate the contrast in the image, but EELS and XPS both found that the total Si content was only 1–2 at.%

More at Oak Ridge National Lab (Fig. 1.10). In literature, TEM images of GO have been widely reported [74]. People observe large domain of $sp^2$ carbons and attached oxygenated moieties. It is also worthwhile to mention the Si contamination is often observed in high-resolution microscopic images of GO. Although the synthesis process of GO doesn't involve any Si-containing compound, we found that Si comes from the deionized water (>18 M$\Omega$) generated by a Millipore filter, which contains certain amount of $SiO_2$.

**Fig. 1.11** (**a**) A tapping
mode AFM image of
graphene oxide (GO) sheets
on mica surface, (**b**) the
height profile of the AFM
image (Reproduced from ref.
[75] with permission of the
Royal Society of Chemistry)

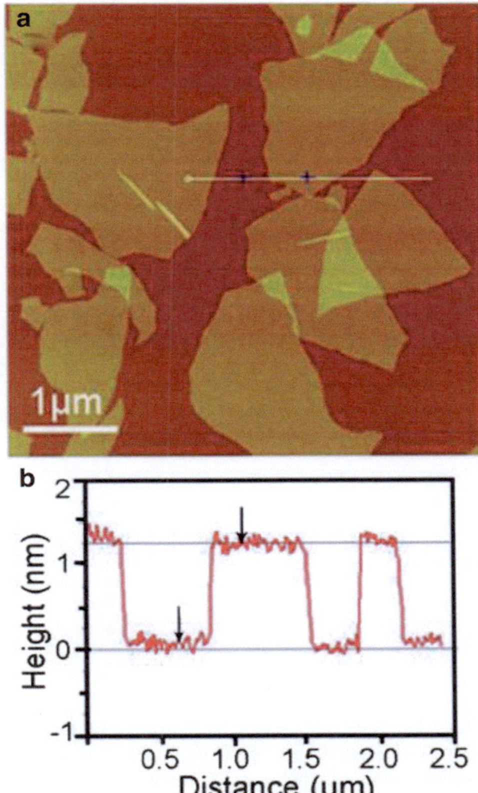

## 1.4.2  *Atomic Force Microscopy (AFM) and Scanning Tunneling Microscopy (STM)*

Atomic force microscopy (AFM) is usually used to measure the thickness and characterize the lateral morphologies of GO sheets, where the spin coating of GO onto a flat surface (typically a Si wafer) is required. People often use piranha solution to pretreat the Si surface, rendering it super-hydrophilic, in order to facilitate the GO coating. Figure 1.11 (adapted from ref [75]) showed a typical AFM image of pristine GO sheets on mica surface. As clearly outlined, the thickness of these flakes was measured to be ~1.2 nm. Researchers tend to agree with the explanation of its higher thickness, as compared to the 3.4 Å of graphene, by the existence of three layers of atoms from the oxygenated functional groups attached to both sides of the GO basal planes. While the statement is true regarding GO structure, we think the resolution of AFM along the vertical direction plays an important role here. Although up to 1 Å resolution has been claimed by AFM [76], we found it hard to get very accurate or solid measurement data when the sample thickness falls below 1 nm. Having that said, seldom reports show the decrease of thickness in GO after reduction by AFM measurements [77].

                                                                                    

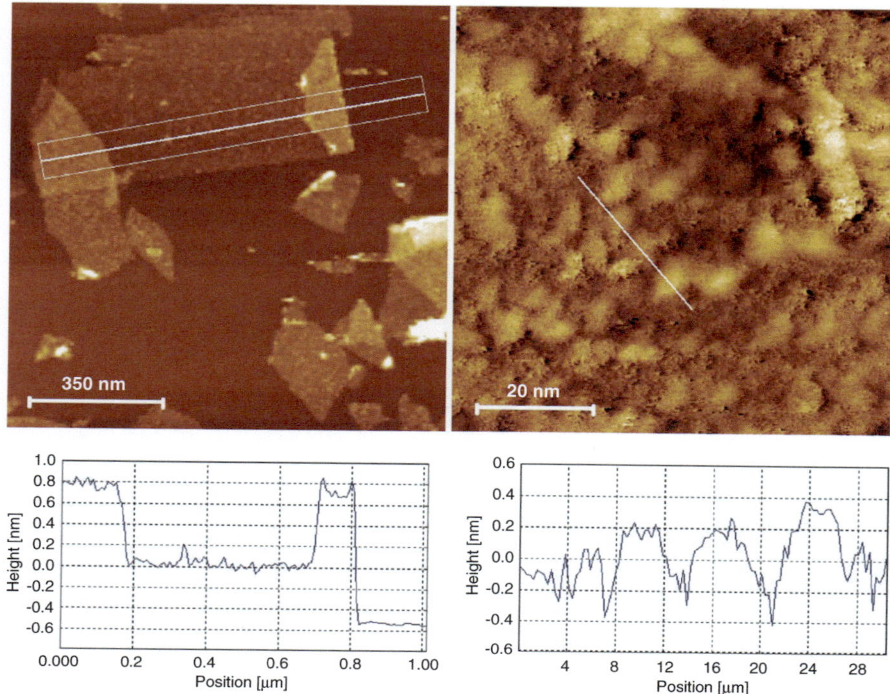

**Fig. 1.12** Nanometer-scale STM images (**a**, **c**) and line profiles (**b**, **d**) of chemically reduced graphene oxide nanosheets deposited onto HOPG from their dispersion in water. The line profile in panel **b** represents the average obtained within the marked *rectangle* in panel **a**, whereas that in **d** was taken along the *white line* marked in panel **c**. Tunneling parameters: $I=0.5$ nA, $V=100$ mV (**a**); $I=0.5$ nA, $V=500$ mV (**c**) (Reprinted with permission from ref. [46]. Copyright (2009) American Chemical Society)

On the other hand, due to the insulating nature of GO, scanning tunneling microscopy (STM) characterizations of GO are rarely reported [46]. STM images on reduced GO (rGOs) which is much more electronically conductive are shown in Fig. 1.12 (adapted from ref [46]). For Fig. 1.12 c, an STM image at a higher resolution, the surface morphology was really random, no periodicity can be observed, and it's definitely hard to correlate it with the high-resolution TEM images we showed earlier (Fig. 1.10). This is mainly why in literature, AFM and STM are only used to emphasize the thickness information for GO and rGO samples.

### 1.4.3  Thermogravimetric Analysis (TGA) and Differential Scanning Calorimetry (DSC)

TGA offers information on thermal stability of GO. Unfortunately GO is thermally unstable; when heated in Ar or $N_2$, GO starts decomposing slowly above 60–80 °C and loses up to 60 % of its total weight when heated up to 950 °C. The loss of

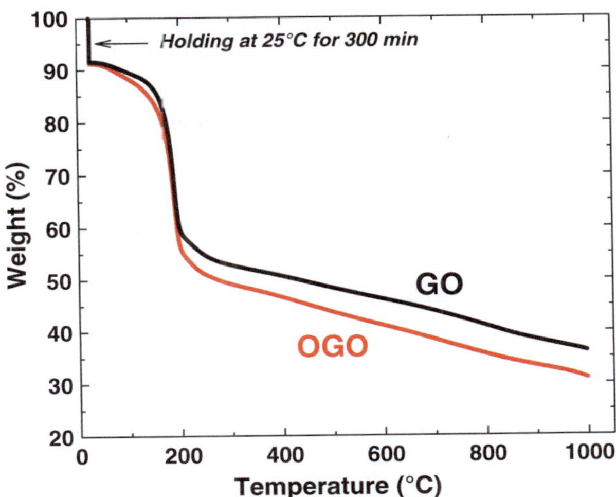



**Fig. 1.13** Thermal gravimetric analysis (TGA) of freestanding GO and ozonated GO (OGO) films. Ramping rate: 2 °C/min in ultrahigh purity N2. Both samples are kept in the N2 atmosphere at 25 °C from 300 min before being heated to higher temperatures to remove all physically absorbed water (Adapted with permission from ref. [71]. Copyright (2014) John Wiley & Sons, Inc.)

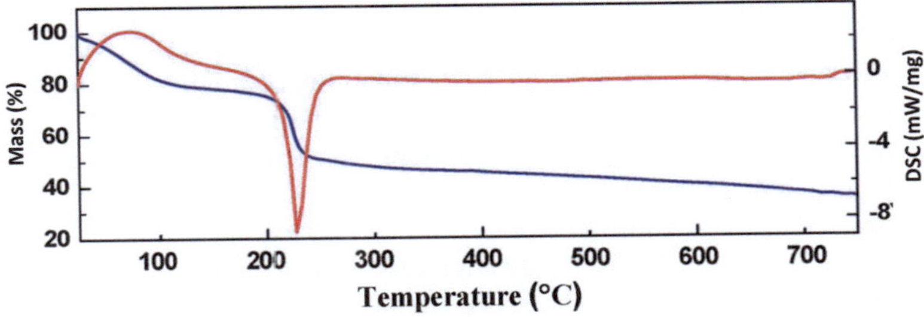

**Fig. 1.14** TG–DSC curves of graphene oxide (Reproduced from ref. [78] with permission of the Royal Society of Chemistry)

adsorbed water will also occur during this process. It is important to mention that, for GO TGA analysis, the temperature ramping rate is usually set to be very low (around 1 °C/min), in case the rapidly evolved gases explode and cause problems. In Fig. 1.13, authors compared the TGA behavior of GO and ozonated GO (OGO) samples, where OGO showed a slightly higher weight loss around 200 °C.

DSC of GO is not often reported, since not much information can be provided. Here we shown a combination of TG and DSC analysis on GO (Fig. 1.14). The experiment was performed under a $N_2$ atmosphere with a heating rate of 10 °C/min. Since they didn't dry out the sample before heating, the first endothermal peak in DSC is due to water desorption, around 17 wt% in their case. The major mass loss occurs at

200–240 °C for their samples. Correspondingly, the DSC curve shows a strong exothermal peak centered at ca. 226 °C. It can be attributed to the decomposition of the labile oxygen-containing moieties. It is worthwhile to note the higher decomposition temperature observed in their case as compared to Fig. 1.13, which can be marginally explained by the higher ramping rate in their case, different structures of GO, as well as the differences in atmospheric pressure.

## 1.5 Concluding Remarks

GO, a group of gigantic organic molecules first discovered over one and a half centuries ago, has reemerged as an important precursor to graphene since the last decade, yet now has written its own history independent of its relationship with graphene.

In this chapter, we summarize and discuss its synthesis, characterizations, structures, and general characterizations based on the fundamental understanding in chemistry and materials science. Starting from natural graphitic material, GO can be prepared in a large amount by strong chemical oxidation. Various characterizations have proved GO to be an oxidized carbon compound with two-dimensional structure, in other words a sheet of fused hexagonal rings with lots of oxygenated groups on both sides. The mixture of $sp^2$ and $sp^3$ carbon atoms in GO makes it corrugated, amphiphilic, and fluorescent, while the oxygen-containing functionalization renders its hygroscopicity, dispersibility, and chemical reactivity. Its two-dimensional nature leads to large-scale availability of atomically thin, transparent GO films, which can be further reduced to form transparent, conductive membranes. Wide applications extending into more than ten different categories have been demonstrated, and GO has become an important material in both fundamental science and contemporary technologies.

The salient feature in GO has made it hard to identify its precise chemical structure. Although many researchers have tried to tailor its molecular structure by various chemicals, the well-resolved control or manipulation of its $sp^2/sp^3$ domain; the location, density, and types of chemical functionalizations; and, of course, the size, shape, and edge structure of GO itself remain far from realization. The development of better chemical processes to eliminate the oxygenated groups and to restore $\pi$ conjugation is also of interest for technological applications. Water-stable cross-linking strategy would also be intriguing in terms of enhanced mechanical properties and wider technological applications.

**Acknowledgment** W. G. sincerely thank for the start-up funding support from the Department of Textile Engineering, Chemistry & Science at North Carolina State University, Raleigh, NC.

# References


1. Coleman JN, Lotya M, O'Neill A, Bergin SD, King PJ, Khan U, Young K, Gaucher A, De S, Smith RJ, Shvets IV, Arora SK, Stanton G, Kim H-Y, Lee K, Kim GT, Duesberg GS, Hallam T, Boland JJ, Wang JJ, Donegan JF, Grunlan JC, Moriarty G, Shmeliov A, Nicholls RJ, Perkins JM, Grieveson EM, Theuwissen K, McComb DW, Nellist PD, Nicolosi V (2011) Two-dimensional nanosheets produced by liquid exfoliation of layered materials. Science 331: 568–571
2. Novoselov KS, Geim AK, Morozov SV, Jiang D, Katsnelson MI, Grigorieva IV, Dubonos SV, Firsov AA (2005) Two-dimensional gas of massless Dirac fermions in graphene. Nature 438:197–200
3. Novoselov KS, Geim AK, Morozov SV, Jiang D, Zhang Y, Dubonos SV, Grigorieva IV, Firsov AA (2004) Electric field effect in atomically thin carbon films. Science 306:666–669
4. Viculis LM, Mack JJ, Kaner RB (2003) A chemical route to carbon nanoscrolls. Science 299:1361
5. Hernandez Y, Nicolosi V, Lotya M, Blighe FM, Sun Z, De S, McGovern IT, Holland B, Byrne M, Gun'Ko YK, Boland JJ, Niraj P, Duesberg G, Krishnamurthy S, Goodhue R, Hutchison J, Scardaci V, Ferrari AC, Coleman JN (2008) High-yield production of graphene by liquid-phase exfoliation of graphite. Nat Nanotechnol 3:563–568
6. Li X, Cai W, An J, Kim S, Nah J, Yang D, Piner R, Velamakanni A, Jung I, Tutuc E, Banerjee SK, Colombo L, Ruoff RS (2009) Large-area synthesis of high-quality and uniform graphene films on copper foils. Science 324:1312–1314
7. Berger C, Song Z, Li X, Wu X, Brown N, Naud C, Mayou D, Li T, Hass J, Marchenkov AN, Conrad EH, First PN, de Heer WA (2006) Electronic confinement and coherence in patterned epitaxial graphene. Science 312:1191–1196
8. Cai J, Ruffieux P, Jaafar R, Bieri M, Braun T, Blankenburg S, Muoth M, Seitsonen AP, Saleh M, Feng X, Mullen K, Fasel R (2010) Atomically precise bottom-up fabrication of graphene nanoribbons. Nature 466:470–473
9. Tomović Ž, Watson MD, Müllen K (2004) Superphenalene-based columnar liquid crystals. Angew Chem Int Ed 43:755–758
10. Treier M, Pignedoli CA, Laino T, Rieger R, Mullen K, Passerone D, Fasel R (2011) Surface-assisted cyclodehydrogenation provides a synthetic route towards easily processable and chemically tailored nanographenes. Nat Chem 3:61–67
11. Gilje S, Han S, Wang M, Wang KL, Kaner RB (2007) A chemical route to graphene for device applications. Nano Lett 7:3394–3398
12. Stankovich S, Dikin DA, Piner RD, Kohlhaas KA, Kleinhammes A, Jia Y, Wu Y, Nguyen ST, Ruoff RS (2007) Synthesis of graphene-based nanosheets via chemical reduction of exfoliated graphite oxide. Carbon 45:1558–1565
13. Brodie BC (1859) On the atomic weight of graphite. Philos Trans R Soc Lond B Biol Sci 149:249–259
14. Boukhvalov DW, Katsnelson MI (2008) Modeling of graphite oxide. J Am Chem Soc 130:10697–10701
15. Hummers WS, Offeman RE (1958) Preparation of graphitic oxide. J Am Chem Soc 80:1339
16. Marcano DC, Kosynkin DV, Berlin JM, Sinitskii A, Sun Z, Slesarev A, Alemany LB, Lu W, Tour JM (2010) Improved synthesis of graphene oxide. ACS Nano 4:4806–4814
17. Cai W, Piner RD, Stadermann FJ, Park S, Shaibat MA, Ishii Y, Yang D, Velamakanni A, An SJ, Stoller M, An J, Chen D, Ruoff RS (2008) Synthesis and solid-state NMR structural characterization of 13C-labeled graphite oxide. Science 321:1815–1817
18. Gao W, Alemany LB, Ci L, Ajayan PM (2009) New insights into the structure and reduction of graphite oxide. Nat Chem 1:403–408
19. Mattevi C, Eda G, Agnoli S, Miller S, Mkhoyan KA, Celik O, Mostrogiovanni D, Granozzi G, Garfunkel E, Chhowalla M (2009) Evolution of electrical, chemical, and structural properties

of transparent and conducting chemically derived graphene thin films. Adv Funct Mater 19:2577–2583

20. Staudenmaier L (1898) Verfahren zur Darstellung der Graphitsäure. Berichte der deutschen chemischen Gesellschaft 31:1481–1487
21. Staudenmaier L (1899) Verfahren zur Darstellung der Graphitsäure. Berichte der deutschen chemischen Gesellschaft 32:1394–1399
22. Hofmann U, Frenzel A (1934) The reduction of graphite oxide with hydrogen sulphide. Kolloid-Zeitschrift 68:149–151
23. Hofmann U, Holst R (1939) The acidic nature and the methylation of graphitoxide. Berichte der deutschen chemischen Gesellschaft 72:754–771
24. Chua CK, Sofer Z, Pumera M (2012) Graphite oxides: effects of permanganate and chlorate oxidants on the oxygen composition. Chemistry 18:13453–13459
25. Dreyer DR, Todd AD, Bielawski CW (2014) Harnessing the chemistry of graphene oxide. Chem Soc Rev 43:5288–5301
26. Kovtyukhova NI, Ollivier PJ, Martin BR, Mallouk TE, Chizhik SA, Buzaneva EV, Gorchinskiy AD (1999) Layer-by-layer assembly of ultrathin composite films from micron-sized graphite oxide sheets and polycations. Chem Mater 11:771–778
27. Chen J, Yao B, Li C, Shi G (2013) An improved Hummers method for eco-friendly synthesis of graphene oxide. Carbon 64:225–229
28. Chen Z-L, Kam F-Y, Goh RG, Song J, Lim G-K, Chua L-L (2013) Influence of graphite source on chemical oxidative reactivity. Chem Mater 25:2944–2949
29. Botas C, Álvarez P, Blanco C, Santamaría R, Granda M, Ares P, Rodríguez-Reinoso F, Menéndez R (2012) The effect of the parent graphite on the structure of graphene oxide. Carbon 50:275–282
30. Kim JE, Han TH, Lee SH, Kim JY, Ahn CW, Yun JM, Kim SO (2011) Graphene oxide liquid crystals. Angew Chem Int Ed 50:3043–3047
31. Luo J, Cote LJ, Tung VC, Tan ATL, Goins PE, Wu J, Huang J (2010) Graphene oxide nanocolloids. J Am Chem Soc 132:17667–17669
32. Rosca ID, Watari F, Uo M, Akasaka T (2005) Oxidation of multiwalled carbon nanotubes by nitric acid. Carbon 43:3124–3131
33. Becker L, Poreda RJ, Bunch TE (2000) Fullerenes: an extraterrestrial carbon carrier phase for noble gases. Proc Natl Acad Sci U S A 97:2979–2983
34. McCleverty JA (1989) Advanced inorganic-chemistry, 5th edition—Cotton, FA, Wilkinson, G. Nature 338:182
35. Koch KR (1982) Oxidation by $Mn_2O_7$: An impressive demonstration of the powerful oxidizing property of dimanganeseheptoxide. J Chem Educ 59:973
36. Simon A, Dronskowski R, Krebs B, Hettich B (1987) The crystal structure of $Mn_2O_7$. Angew Chem Int Ed Engl 26:139–140
37. Higginbotham AL, Kosynkin DV, Sinitskii A, Sun Z, Tour JM (2010) Lower-defect graphene oxide nanoribbons from multiwalled carbon nanotubes. ACS Nano 4:2059–2069
38. Dimiev A, Kosynkin DV, Alemany LB, Chaguine P, Tour JM (2012) Pristine graphite oxide. J Am Chem Soc 134:2815–2822
39. Kim F, Luo J, Cruz-Silva R, Cote LJ, Sohn K, Huang J (2010) Self-propagating domino-like reactions in oxidized graphite. Adv Funct Mater 20:2867–2873
40. Lerf A, He HY, Forster M, Klinowski J (1998) Structure of graphite oxide revisited. J Phys Chem B 102:4477–4482
41. Casabianca LB, Shaibat MA, Cai WW, Park S, Piner R, Ruoff RS, Ishii Y (2010) NMR-based structural modeling of graphite oxide using multidimensional 13C solid-state NMR and ab initio chemical shift calculations. J Am Chem Soc 132:5672–5676
42. Szabo T, Berkesi O, Dekany I (2005) DRIFT study of deuterium-exchanged graphite oxide. Carbon 43:3186–3189
43. Szabo T, Berkesi O, Forgo P, Josepovits K, Sanakis Y, Petridis D, Dekany I (2006) Evolution of surface functional groups in a series of progressively oxidized graphite oxides. Chem Mater 18:2740–2749

44. Kudin KN, Ozbas B, Schniepp HC, Prud'homme RK, Aksay IA, Car R (2008) Raman spectra of graphite oxide and functionalized graphene sheets. Nano Lett 8:36–41

45. Ishigami M, Chen JH, Cullen WG, Fuhrer MS, Williams ED (2007) Atomic structure of graphene on SiO2. Nano Lett 7:1643–1648

46. Paredes JI, Villar-Rodil S, Solis-Fernandez P, Martinez-Alonso A, Tascon JMD (2009) Atomic force and scanning tunneling microscopy imaging of graphene nanosheets derived from graphite oxide. Langmuir 25:5957–5968

47. Gomez-Navarro C, Meyer JC, Sundaram RS, Chuvilin A, Kurasch S, Burghard M, Kern K, Kaiser U (2010) Atomic structure of reduced graphene oxide. Nano Lett 10:1144–1148

48. Wilson NR, Pandey PA, Beanland R, Young RJ, Kinloch IA, Gong L, Liu Z, Suenaga K, Rourke JP, York SJ, Sloan J (2009) Graphene oxide: structural analysis and application as a highly transparent support for electron microscopy. ACS Nano 3:2547–2556

49. Eda G, Mattevi C, Yamaguchi H, Kim H, Chhowalla M (2009) Insulator to semimetal transition in graphene oxide. J Phys Chem C 113:15768–15771

50. Kaiser AB, Gomez-Navarro C, Sundaram RS, Burghard M, Kern K (2009) Electrical conduction mechanism in chemically derived graphene monolayers. Nano Lett 9:1787–1792

51. Luo ZT, Lu Y, Somers LA, Johnson ATC (2009) High Yield Preparation of Macroscopic Graphene Oxide Membranes. J Am Chem Soc 131(3):898–899

52. Ruess G (1947) Über das Graphitoxyhydroxyd (Graphitoxyd). Monatsch Chem 76:381–417

53. Mermoux M, Chabre Y, Rousseau A (1991) FTIR and C-13 NMR-study of graphite oxide. Carbon 29:469–474

54. Boehm HP, Clauss A, Hofmann U, Fischer GO (1962) Dunnste Kohlenstoff-Folien. Z Naturforschg 17:150–157

55. Scholz W, Boehm HP (1969) Graphite oxide. 6. Structure of graphite oxide. Z Anorg Allg Chem 369:327–340

56. Nakajima T, Mabuchi A, Hagiwara R (1988) A new structure model of graphite oxide. Carbon 26:357–361

57. He HY, Riedl T, Lerf A, Klinowski J (1996) Solid-state NMR studies of the structure of graphite oxide. J Phys Chem 100:19954–19958

58. Lerf A, He HY, Riedl T, Forster M, Klinowski J (1997) C-13 and H-1 MAS NMR studies of graphite oxide and its chemically modified derivatives. Solid State Ion 101:857–862

59. He H, Klinowski J, Forster M, Lerf A (1998) A new structural model for graphite oxide. Chem Phys Lett 287:53–56

60. Dimiev AM, Alemany LB, Tour JM (2012) Graphene oxide. Origin of acidity, its instability in water, and a new dynamic structural model. ACS Nano 7:576–588

61. Kim S, Zhou S, Hu Y, Acik M, Chabal YJ, Berger C, de Heer W, Bongiorno A, Riedo E (2012) Room-temperature metastability of multilayer graphene oxide films. Nat Mater 11:544–549

62. Kumar P, Maiti UN, Lee KE, Kim SO (2014) Rheological properties of graphene oxide liquid crystal. Carbon 80:453–461

63. Rourke JP, Pandey PA, Moore JJ, Bates M, Kinloch IA, Young RJ, Wilson NR (2011) The real graphene oxide revealed: stripping the oxidative debris from the graphene-like sheets. Angew Chem Int Ed 50:3173–3177

64. Guo Z, Wang S, Wang G, Niu Z, Yang J, Wu W (2014) Effect of oxidation debris on spectroscopic and macroscopic properties of graphene oxide. Carbon 76:203–211

65. Thomas HR, Day SP, Woodruff WE, Vallés C, Young RJ, Kinloch IA, Morley GW, Hanna JV, Wilson NR, Rourke JP (2013) Deoxygenation of graphene oxide: reduction or cleaning? Chem Mater 25:3580–3588

66. Liu L, Wang L, Gao J, Zhao J, Gao X, Chen Z (2012) Amorphous structural models for graphene oxides. Carbon 50:1690–1698

67. Eigler S, Dotzer C, Hof F, Bauer W, Hirsch A (2013) Sulfur species in graphene oxide. Chem-Eur J 19:9490–9496

68. Zhou S, Bongiorno A (2013) Chemical stability of epoxy functionalizations of graphene: a density functional theory study. MRS Online Proc Libr 1549:19–24

69. Kumar PV, Bardhan NM, Tongay S, Wu J, Belcher AM, Grossman JC (2014) Scalable enhancement of graphene oxide properties by thermally driven phase transformation. Nat Chem 6:151–158
70. Zhou S, Bongiorno A (2013) Origin of the chemical and kinetic stability of graphene oxide. Sci Rep 3
71. Gao W, Wu G, Janicke MT, Cullen DA, Mukundan R, Baldwin JK, Brosha EL, Galande C, Ajayan PM, More KL (2014) Ozonated graphene oxide film as a proton-exchange membrane. Angew Chem Int Ed 53:3588–3593
72. Chua CK, Pumera M (2014) Light and atmosphere affect the quasi-equilibrium states of graphite oxide and graphene oxide powders. Small
73. Mkhoyan KA, Contryman AW, Silcox J, Stewart DA, Eda G, Mattevi C, Miller S, Chhowalla M (2009) Atomic and electronic structure of graphene-oxide. Nano Lett 9:1058–1063
74. Pacilé D, Meyer J, Fraile Rodríguez A, Papagno M, Gomez-Navarro C, Sundaram R, Burghard M, Kern K, Carbone C, Kaiser U (2011) Electronic properties and atomic structure of graphene oxide membranes. Carbon 49:966–972
75. Zhang JL, Yang HJ, Shen GX, Cheng P, Zhang JY, Guo SW (2010) Reduction of graphene oxide via L-ascorbic acid. Chem Commun 46:1112–1114
76. Alexander S, Hellemans L, Marti O, Schneir J, Elings V, Hansma P, Longmire M, Gurley J (1989) An atomic-resolution atomic-force microscope implemented using an optical lever. J Appl Phys 65:164–167
77. Ma C, Chen Z, Fang M, Lu H (2012) Controlled synthesis of graphene sheets with tunable sizes by hydrothermal cutting. J Nanopart Res 14:1–9
78. Ji Z, Zhu G, Shen X, Zhou H, Wu C, Wang M (2012) Reduced graphene oxide supported FePt alloy nanoparticles with high electrocatalytic performance for methanol oxidation. New J Chem 36:1774–1780

# Chapter 2
# Spectroscopy and Microscopy of Graphene Oxide and Reduced Graphene Oxide

Matthew P. McDonald, Yurii Morozov, Jose H. Hodak, and Masaru Kuno

**Abstract** Graphene oxide (GO) is an important material that provides a scalable approach for obtaining chemically derived graphene. Its optical and electrical properties are largely determined by the presence of oxygen-containing functionalities, which decorate its basal plane. This chemical derivatization results in useful properties such as the existence of a band gap as well as emission spanning both the visible and near infrared. Notably, GO's optical and electrical properties can be altered through reduction, which proceeds through the removal of these oxygen-containing functional groups. However, widely variable behavior has been observed regarding the evolution of GO's optical response during reduction. These discrepancies arise from the different reduction methods being used and, in part, from the fact that nearly all prior measurements have been ensemble studies. Consequently, detailed mechanistic studies of GO reduction are needed which can transcend the limitations of ensemble averaging.

In this chapter, we show the spectroscopic evolution of GO's optical properties during photoreduction at the single-sheet level. Laser-induced reduction, in particular, offers a unique and potentially controllable method for producing reduced GO (rGO), a material with properties similar to those of graphene. However, given the complexity of GO's photoreduction mechanism, microscopic monitoring of the process is essential to understanding and ultimately exploiting this approach.

**Keywords** Graphene oxide • Reduced graphene oxide • Photolysis • Reduction • Photobrightening • Absorption • Emission • Absorption coefficient • Fluorescence intermittency

---

M.P. McDonald • M. Kuno (✉)
Department of Chemistry and Biochemistry, University of Notre Dame,
Notre Dame, IN 46556, USA
e-mail: mkuno@nd.edu

Y. Morozov
Department of Physics, Taras Shevchenko National University of Kiev, Kiev, Ukraine

Department of Chemistry and Biochemistry, University of Notre Dame,
Notre Dame, IN 46556, USA

J.H. Hodak
INQUIMAE—Departamento de Química Inorgánica, Analítica y Química Físca, Facultad de Ciencias Exactas y Naturales, University of Buenos Aires, Buenos Aires, Argentina



W. Gao (ed.), *Graphene Oxide*, DOI 10.1007/978-3-319-15500-5_2

## 2.1  Graphene, GO, and rGO

### 2.1.1  Physical Properties of Graphene, GO, and rGO

Since its discovery in the early 2000s [1], graphene has emerged as a major source for exciting new physics and applications. This atomically thin, hexagonal lattice of $sp^2$ hybridized carbons exhibits many exceptional properties, including optical transparency [2], large carrier mobilities [3], and excellent electrical [4] as well as thermal [5] conductivities. Such properties have led to its use in a variety of applications, such as in transparent/flexible electrodes [6], and in field-effect transistors (FETs) [7].

Many graphene production methods have emerged in recent years [1, 8, 9]. However, for the most part, they remain costly and time consuming. This limits graphene's practical usage. A scalable, low-cost approach for obtaining mass quantities of high-quality graphene is therefore needed.

The chemical oxidation of graphite and its subsequent exfoliation in water result in single-layered GO sheets. In this regard, GO is a promising precursor for chemically derived graphene [10]. Though originally synthesized in 1859 by Brodie [11], a safer, more reliable method for producing graphite oxide was only developed in the 1950s by Hummers and Offeman [12]. Modern methods are essentially slightly modified versions of this synthesis and include a sonication step to exfoliate graphite oxide [10]. Figure 2.1 provides a simplified depiction of how GO is produced today [10], with Chap. 1 of this book providing a more thorough discussion of GO production methods.

Physically, graphene oxide is an atomically thin, semi-aromatic network of $sp^2/sp^3$ bonded carbon atoms intermittently decorated with oxygen-containing functionalities [13]. These functional groups include hydroxyl (OH), epoxy (C–O–C), carbonyl (C=O), and carboxyl (COOH) species, with OH and C–O–C being the dominant groups across GO's basal plane [13]. Their subsequent removal through disproportionation reactions [14] results in rGO, a chemical analogue of graphene.

GO reduction therefore provides a scalable, low-cost approach for obtaining a graphene-like material [10]. In this regard, reduction can be achieved through various means. They include chemical [15], thermal [16], photothermal [17], and laser-induced [18] reduction methods. All remove GO's oxygen-containing functionalities and simultaneously convert basal plane carbon atoms from $sp^3$ to $sp^2$ hybridization, restoring the system's aromaticity to a certain extent.

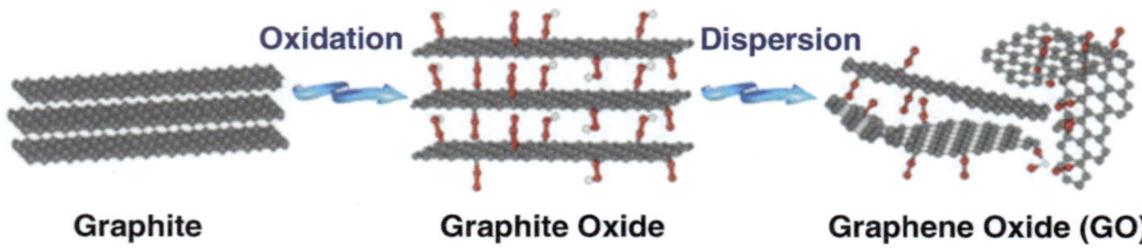


**Fig. 2.1** Cartoon illustrating the chemical production of graphene oxide sheets. Reprinted with permission from Zhang Y.L., et al., *Adv. Optical Mater.* **2014**, *2*, 10-28. Copyright 2014 Wiley-VCH

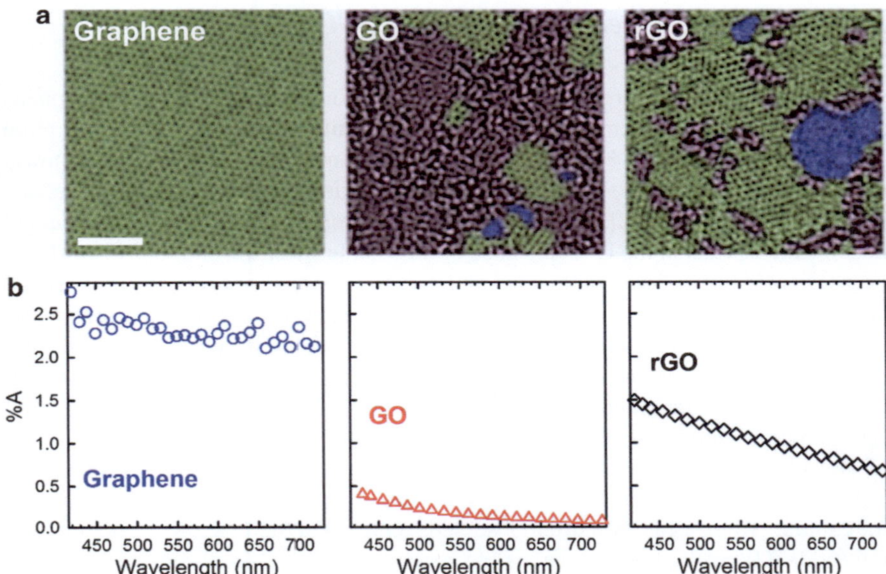


**Fig. 2.2** (**a**) Aberration-corrected TEM images of graphene, GO, and rGO. *Green regions* indicate perfect $sp^2$ character, *red regions* denote disordered (oxidized) portions, and *blue regions* depict defects/holes. *Scale bar*: 2 nm. Adapted with permission from Erickson K., et al., *Adv. Mater.* 2010, *22*, 4467-4472. Copyright 2010 Wiley-VCH. (**b**) Visible absorption spectra of single-layer graphene (*blue circles*; adapted from ref. [2], single-layer GO (*red triangles*; adapted from ref. [21]), and single-layer rGO (*black diamonds*; adapted from ref. [21])

To better elucidate the mechanism behind GO to rGO interconversion for possible large-scale production, significant structural characterization has been conducted on these materials at the microscopic level [19, 20]. Electron and tunneling microscopies, in particular, have provided detailed insight into GO's chemical and physical structure. As an illustration, Fig. 2.2a shows high-contrast aberration-corrected transmission electron microscopy (TEM) images of pristine graphene, GO, and rGO [19]. In these micrographs, $sp^2$ regions are colored green, disordered (oxidized) $sp^3$ regions are shown in red, and defects/holes are highlighted in blue [19]. These images show that GO clearly regains much of its $sp^2$ character upon reduction and begins to resemble graphene. However, oxidized regions still exist within rGO and are accompanied by the presence of holes and defects.

These TEM measurements [19]—along with scanning tunneling microscopy (STM) [20] experiments—also show sizable oxygen functionality and $sp^2$ size distribution heterogeneities within GO and rGO. GO/rGO's oxygen-containing functionalities are additionally observed to be dynamic. Namely, under electron beam irradiation, they migrate across GO/rGO's basal plane [19]. This, coupled with rGO's imperfect/defective structure, highlights the complicated chemistry underlying GO-to-rGO interconversion. Consequently, detailed studies of GO's deoxygenation mechanisms are needed to fully realize its immense potential as a graphene precursor.

## *2.1.2   Optical Properties of Graphene, GO, and rGO*

GO's optical properties are significantly different from those of graphene. Graphene is a zero-gap semiconductor, which possesses a significant, near-constant absorption across the visible (~2.3 % absorption, %A, first panel, Fig. 2.2b) [2]. By contrast, GO's optical absorption is nearly an order of magnitude smaller (~0.3 %A, second panel, Fig. 2.2b) [21, 22]. Furthermore, GO's absorption spectrum is wavelength dependent; it peaks in the UV/blue edge of the visible spectrum and falls towards the near infrared (NIR). Discernible $\pi$–$\pi^*$ and n–$\pi^*$ transitions are also present in its UV absorption (200–320 nm) and arise from the presence of oxygen-containing functionalities [23, 24]. GO's optical response is thus exquisitely sensitive to its degree of oxidation [25]. As an illustration, the last two panels in Fig. 2.2b show that GO's absorption dramatically increases across the visible upon the removal of these functionalities and approaches graphene's 2.3 %A limit [21, 26].

An additional difference from graphene is that, upon absorbing UV or visible light, GO fluoresces [23, 24, 27–30]. Two broad (200–300 nm full width at half maximum, fwhm) peaks are often observed, centered in the blue (400–500 nm) and red (600–700 nm) regions of the visible spectrum [27, 31]. The origin of this emission is debated. Conventional wisdom suggests that it arises from carrier recombination in isolated/confined $sp^2$ clusters where varying domain sizes lead to a distribution of energy gaps that results in a spectrally broad emission [31]. However, quasi-molecular transitions and even heterogeneous carrier relaxation kinetics have been invoked to explain its origin and appearance [28, 29]. In the former case, intriguing evidence for quasi-molecular photoluminescence, arising from basal plane oxygen-functionalized aromatic clusters, has been observed in GO excitation spectra [29]. In the latter case, optical Kerr-gate photoluminescence measurements have revealed clear spectral dynamics following excitation. Specifically, GO's emission redshifts from 460 to 690 nm over the course of ~10 ps following excitation and clearly indicates the existence of heterogeneous carrier relaxation kinetics [28].

Next, GO's emission changes gradually upon reduction. However, as with its native emission, the literature is replete with differences in reported behavior. In this regard, GO/rGO's emission has been reported to blueshift [27], redshift [30], or not shift at all [23] from GO's initial spectrum during reduction. To illustrate, Fig. 2.3a shows the emission spectrum of a GO suspension during Xe arc lamp photoreduction [27]. The initial spectrum (solid red line) exhibits GO's characteristic red (~600 nm) and blue (~500 nm) peaks, with the red portion dominating. As reduction progresses, the red peak disappears while the blue feature begins to dominate, ultimately leading to an effective blueshift of the spectrum.

By contrast, Fig. 2.3b shows seemingly contradictory behavior during GO's *chemical* (i.e., hydrazine) reduction. Namely, its emission spectrum redshifts from ~700 to ~850 nm upon reduction [30]. Adding to the confusion, other studies have shown no discernible shift in GO's emission spectrum upon hydrazine reduction (Fig. 2.3c) [23].

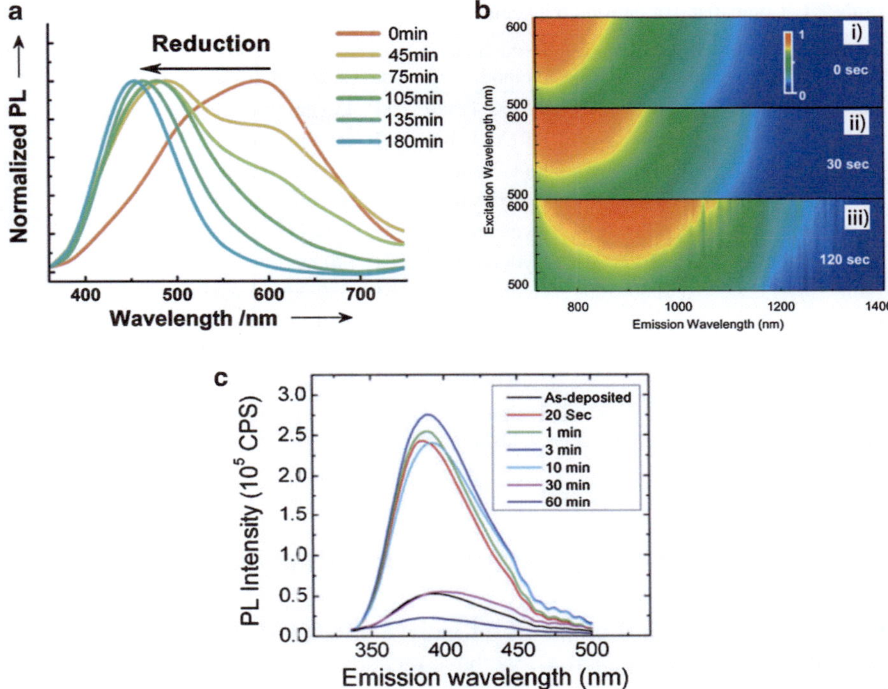


**Fig. 2.3** (**a**) Normalized photoluminescence (PL) spectra of GO suspensions during photoreduction. Reprinted with permission from Chien C-T, et al., *Angew. Chem. Int. Ed.*, 2012, *51*, 6662–6666. Copyright 2012 Wiley-VCH. (**b**) Normalized PL excitation-emission maps of GO during chemical reduction. Reprinted with permission from Luo Z., et al., *Appl. Phys. Lett.*, 2009, *94*, 111909. Copyright 2009 AIP. (**c**) PL spectra of GO thin films during chemical reduction. Reprinted with permission from Eda G., et al., *Adv. Mater.*, 2010, *22*, 505–509. Copyright 2010 Wiley-VCH

Table 2.1 highlights these differences by compiling characteristic features of literature-reported GO/rGO emission spectra observed during reduction. The table is grouped by reduction method. Widely varying responses are seen depending on sample type (suspensions, solid films, or single sheets) and/or reduction method. However, no clear correlation exists between the observed spectral response and reduction method. Additionally, chemical reduction (hydrazine, in particular) appears to contribute most to the variability of the observed behavior.

This variability and lack of consensus stem, in part, from the fact that the vast majority of GO/rGO optical studies have been conducted at the ensemble level [15–17, 23–31, 36]. Consequently, to begin resolving these discrepancies single-GO/rGO sheet optical measurements, which transcend ensemble averaging, are needed to reveal the microscopic photophysics underlying GO-to-rGO interconversion [32, 37].

**Table 2.1** Comparison of GO and rGO emission properties during GO-to-rGO interconversion

| Sample type | $\lambda_{exc}$ (nm) | GO spectrum; max (nm) | Reduction method | rGO spectrum; max | References |
|---|---|---|---|---|---|
| Susp. | 325 | DP; 600 | PT (Xe lamp) | SP; 450 nm (blueshift) | [27] |
| Film | 400 | DP; 690 | PT (Xe lamp) | SP; 460 nm (blueshift) | [28] |
| Sing. | 473 | SP; 700 | L (473 nm) | SP; 625 nm (blueshift) | [32] |
| Sing. | 405 | DP; 690 | L (405 nm) | SP; 500 nm (blueshift) | [33] |
| Film | 488 | SP; 650 | Hydrazine | MP; 700–900 nm (redshift) | [30] |
| Film | 325 | SP; 390 | Hydrazine | SP; 390 nm (no shift) | [23] |
| Film | 325 | MP; 500–700 | Hydrazine | MP; 530 nm (blueshift) | [34] |
| Film | 325 | MP; 500–700 | $T$ (700 °C) | MP; 500–700 nm (slight blueshift) | [34] |
| Susp. | <500 | SP; 650 | $T$ (120 °C) | SP; 475 nm (blueshift) | [35] |

$\lambda_{exc}$ excitation wavelength, *Susp.* suspension, *Sing.* single sheet, *SP* single peak, *DP* double peak, *MP* multiple peaks, *PT* photothermal, *L* laser, *T* thermal

## 2.2 Single-GO Sheet Absorption Microscopy/Spectroscopy

To investigate the optical properties of individual GO and rGO sheets as well as their intrasheet optical properties, we have recently developed a unique toolset to probe both the spatially resolved absorption and emission of individual single-layer GO/rGO specimens [21, 33]. When these measurements are carried out before, during, and after photolytic GO reduction, they provide an unprecedented look at how GO's optical properties evolve throughout its reduction process.

These measurements are part of a broader class of studies we have conducted to probe the photophysics of individual nanostructures [38–42]. Examples include CdSe nanowires [38–42], single-walled carbon nanotubes [42], and Au nanoparticles [42]. Notably, the implementation of single-particle *absorption* microscopy/spectroscopy is unique to our approach and serves to broaden the scope of traditional emission-based single-molecule/particle techniques. In this respect, not all specimens are emissive or possess sizable emission quantum yields (QY). Hence, these materials are inaccessible to current single-particle microscopies.

To acquire single-GO/rGO sheet absorption images and spectra, three diode lasers (405, 520, and 640 nm; Coherent) and a broadband supercontinuum laser (Fianium) are used. Figure 2.4 shows a schematic of the optical setup built to conduct these experiments. The entire system is constructed around an inverted microscope (Nikon) that supports a 3-axis closed-loop piezo positioner (Mad City Labs) and a 2-axis manual micrometer stage (Semprex). GO suspensions are prepared using a modified Hummers synthesis [33]. Optical samples are then made by drop casting dilute suspensions onto fused silica coverslips (Esco) which are then affixed to the piezo stage. This enables precise control of the sample's position. As-prepared samples generally consist of ~90 % single-layer GO sheets [33].

Figure 2.4 shows the optical path of the excitation source. Prior to entering the microscope, each laser is split into signal and reference beams. The signal beam

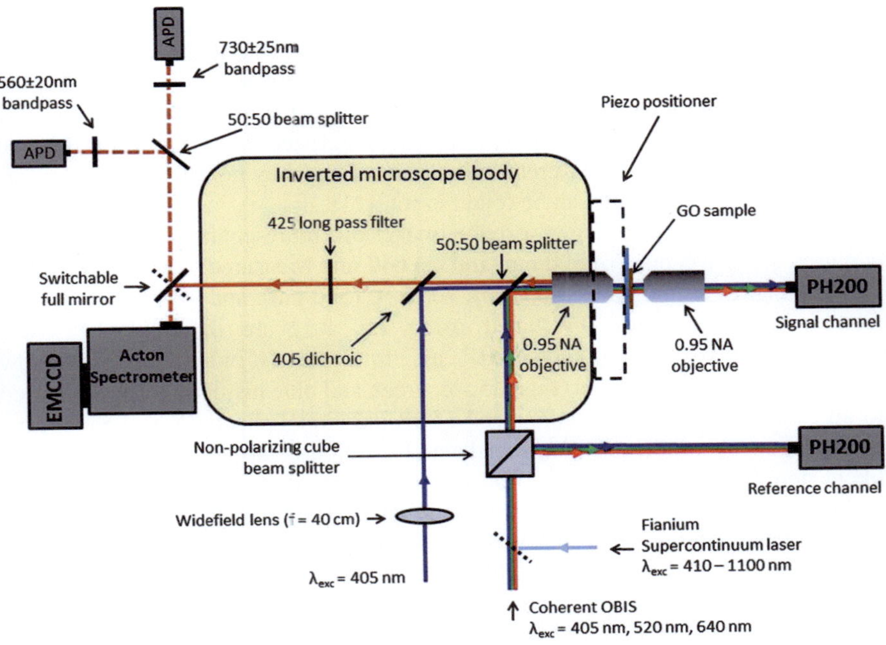


**Fig. 2.4** Optical setup for spatially resolved absorption and emission imaging/spectroscopy

is focused onto samples by way of a high NA microscope objective (Nikon 0.95). The reference beam is simultaneously focused onto a near-shot noise-limited photodetector (Highland Technologies, PH200). Light transmitted through the sample is then collected by a second, collinear, high NA objective (Nikon 0.95) and is focused onto an identical PH200 photodetector. Typical excitation intensities ($I_{exc}$) are $I_{exc}$ ~150 Wcm$^{-2}$. To mitigate laser power fluctuations, the transmitted signal is normalized by the reference detector's voltage.

Spatially resolved absorption images of a single GO/rGO sheet are constructed by raster-scanning the specimen through the probe laser's focus and monitoring the transmitted light intensity ($I$) point by point. The associated incident intensity ($I_0$) is obtained by averaging over a region of the image free of absorbing species. This allows the transmittance ($T = I/I_0$) or % absorption [%A $= (1 - T) \times 100$] to be quantified for every pixel (an approximate 250 nm × 250 nm region). The sample's absorption coefficient ($\alpha$; cm$^{-1}$) is then obtained using [21]

$$\alpha\left(cm^{-1}\right) = \frac{-\ln\left(1 - \dfrac{\%A}{100}\right)}{x\left(cm\right)} \tag{2.1}$$

where we assume a specimen thickness, $x$ (cm), of 1 nm ($1 \times 10^{-7}$ cm) [20]. Typical detection limits for these measurements are ~0.02–0.06 %A and are limited by laser noise, consistent with mode hopping [21].

Corresponding absorption spectra are acquired by dispersing the output of the supercontinuum laser with an acousto-optic tunable filter (AOTF, effective range 410–800 nm). To limit laser exposure during absorption measurements—thereby minimizing photo-induced changes in the specimen—the probe is pulsed once for 3 ms at each point. The transmitted pulse is then digitized, fast Fourier transformed (FFT), and low pass filtered to remove any high-frequency noise. Additional information about the approach can be found in ref. [21].

Figure 2.5 shows resulting absorption images of a representative single GO sheet acquired at (a) 405 nm, (b) 520 nm, and (c) 640 nm. Measured %A-values for this particular sheet are ~0.68 % (405 nm), ~0.25 % (520 nm), and ~0.13 % (640 nm). Corresponding $\alpha$-values (evaluated using Eq. 2.1) are $\alpha_{405} = 6.8 \times 10^4$ cm$^{-1}$, $\alpha_{520} = 2.5 \times 10^4$ cm$^{-1}$, and $\alpha_{640} = 1.3 \times 10^4$ cm$^{-1}$. Interestingly, individual sheets show varying $\alpha$-value distributions (Fig. 2.5a–c; green and blue histograms) as compared to an ensemble distribution composed of 12 individual sheets (Fig. 2.5a–c; gray histograms). This indicates that intersheet absorption heterogeneities exist, which likely stem from sheet-to-sheet variations in their extent of oxidation.

From the images, it is also apparent that there are significant *intra*sheet absorption efficiency differences at each wavelength. Some regions absorb more of a given color than other regions. This is highlighted, for example, by the upper left portion of the sheet profiled in Fig. 2.5a where one sees a large ~1 µm$^2$ region having ~0.9 %A. This value is ~0.2 %A higher than the rest of the sheet. Other regions exhibit varied absorption efficiencies and are readily apparent in Fig. 2.5a–c. An implication of these results is that chemically segregated regions on the ~µm$^2$ scale exist within GO's basal plane. These results are not unique and are observed in all single-layer GO sheets we have studied.

Consequently, extracted %A parameters, averaged over 12 individual GO sheets, are $0.6 \pm 0.1$ %, $0.21 \pm 0.04$ %, and $0.11 \pm 0.03$ % at 405 nm, 520 nm, and 640 nm, respectively. Corresponding $\alpha$-values are $\alpha_{405} = 6 \pm 1 \times 10^4$ cm$^{-1}$, $\alpha_{520} = 2.1 \pm 0.4 \times 10^4$ cm$^{-1}$, and $\alpha_{640} = 1.1 \pm 0.3 \times 10^4$ cm$^{-1}$. To put these numbers into context, graphene possesses an approximate $\alpha$-value of $\alpha = 5.8 \times 10^5$ cm$^{-1}$ across the visible [evaluated using Eq. (2.1) by assuming a thickness of 0.4 nm ($4 \times 10^{-8}$ cm) and %A $= 2.3$ %] [2, 43].

At this point, given that our approach enables the visible spectrum of single-layer GO to be acquired, estimates of GO's full wavelength-dependent absorption coefficient can be made. To illustrate, Fig. 2.5d shows the absorption spectrum of a single GO sheet in green triangles. Average $\alpha$-values from 12 single-sheet absorption images are also shown (blue circles). Superimposed over the single-sheet data is the corresponding ensemble absorption spectrum (dashed red line). Both spectra qualitatively agree, showing a monotonic increase of the absorption as one moves to progressively blue wavelengths. By fitting the single-sheet data, an equation, which provides an estimate for GO's wavelength-dependent $\alpha$-value between $\lambda = 450$–700 nm, is

$$\alpha\left(\text{cm}^{-1}\right) = 665{,}000 - \lambda\left(2{,}970\right) + \lambda^2\left(4.54\right) - \lambda^3\left(2.35 \times 10^{-3}\right). \qquad (2.2)$$

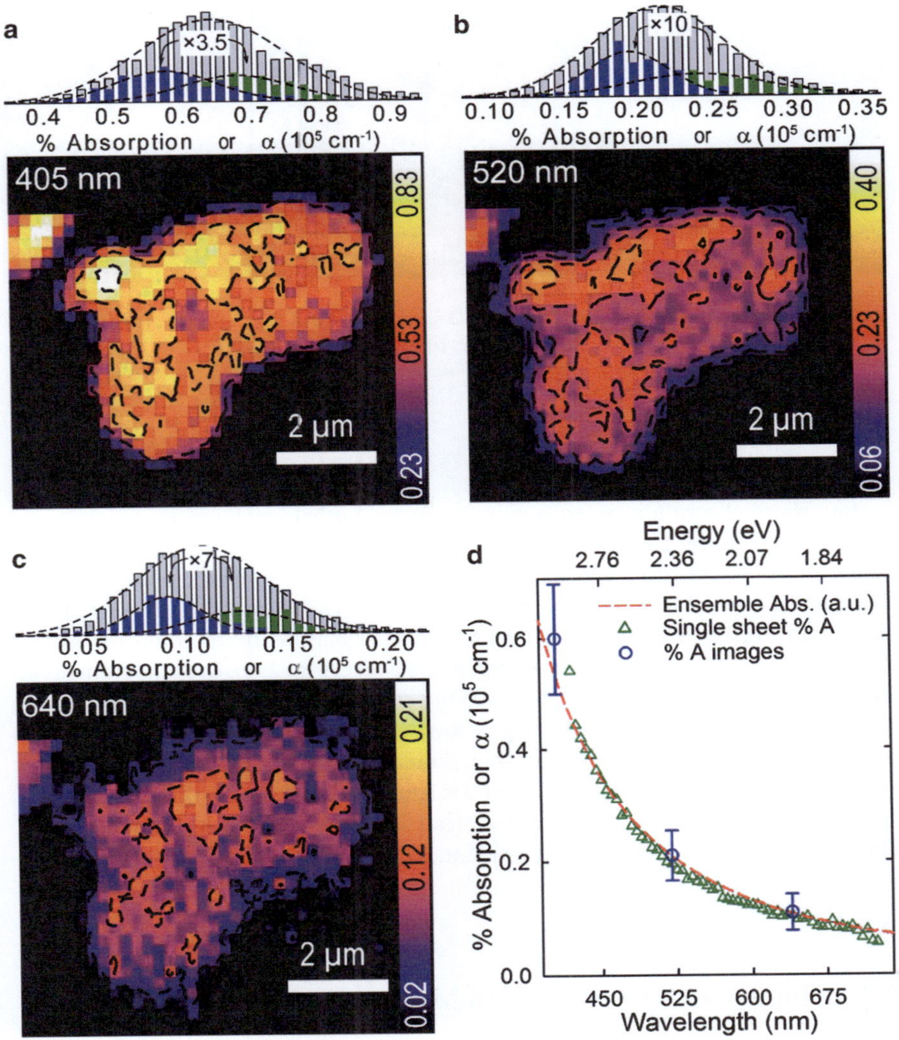



**Fig. 2.5** Absorption images of single-layer GO acquired at (**a**) 405 nm (contour interval, 0.2 % absorption), (**b**) 520 nm (contour interval, 0.05 % absorption), and (**c**) 640 nm (contour interval, 0.05 % absorption). The *false color scale* represents %A. *Gray histograms* above each image show the %A (and corresponding $\alpha$-value) distribution for a 12-sheet GO ensemble. *Green* and *blue* histograms represent the %A distribution for two separate sheets. (**d**) Absorption spectrum for a single GO sheet (*green triangles*) compared to its parent ensemble absorption spectrum (*red dashed line*) and to single $\lambda$ absorption images (*blue circles*). Reprinted with permission from Sokolov, D.A., et al., *Nano Letters* 2014, *14*, 3172–3179. Copyright 2014 American Chemical Society

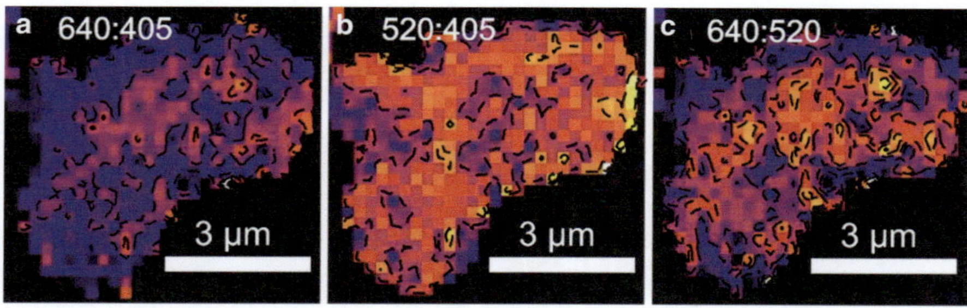



**Fig. 2.6** Absorption ratio maps of the same GO sheet featured in Fig. 2.5. Wavelength ratios are (**a**) 640 nm:405 nm, (**b**) 520 nm:405 nm, and (**c**) 640 nm:520 nm. The *false color scale* represents ratio values, where *yellow/white* is high and *blue/black* is low. Reprinted with permission from Sokolov, D.A., et al., *Nano Letters* 2014, *14*, 3172–3179. Copyright 2014 American Chemical Society

Intrasheet absorption heterogeneities can be further explored by constructing ratios of absorption images obtained at different colors. This, in turn, provides a sense for intrasheet *spectral* heterogeneities present in the sample. Figure 2.6 shows examples of such absorption ratio maps for the same GO sheet featured in Fig. 2.5. In particular, Fig. 2.6a is the resulting ratio image obtained by dividing the sheet's 640 nm image with its 405 nm counterpart. Likewise, Fig. 2.6b, c results from dividing the sheet's 520 nm image with its 405 nm counterpart and its 640 nm image with its 520 nm counterpart. These images show ratio values which are not constant across the specimen. Consequently, they indicate that spectral heterogeneities exist within the sheet. This likely stems from variations in $sp^2$ domain sizes and chemical heterogeneities that favor a disordered electronic energy gap distribution [24, 27]. Ratio maps for other GO sheets show identical intrasheet heterogeneities and can be found in ref. [21].

## 2.3 Single-GO Sheet Emission Microscopy/Spectroscopy

Next, the instrument we have built (Fig. 2.4) supports emission imaging through either standard widefield imaging techniques or confocal scanning microscopy. In both cases, single-layer GO samples are excited with a 405 nm continuous wave (CW) laser diode passed through a high NA objective (Nikon 0.95). The resulting emission can then be studied to more fully elucidate GO's spatial and spectral characteristics.

For widefield imaging, an $f = 40$ cm lens is placed in the excitation path so that the laser comes to a focus just prior to the objective's back focal plane. This results in a large (~30 µm) excitation area with a corresponding excitation intensity of $I_{exc}$ ~300 Wcm$^{-2}$. Emission is collected through the same objective and is passed through a 405 nm dichroic filter (Semrock) and a 425 nm long-pass filter (Chroma).

This light is then focused onto the entrance slit of an imaging spectrometer (Acton) that contains a 50 g/mm (600 nm blaze) grating. An electron-multiplied CCD camera (EMCCD, Andor), positioned at the exit of the spectrometer, acquires both emission images and spectra.

For confocal measurements, the collimated laser is passed directly into the objective and results in a diffraction-limited spot at the focus ($I_{exc} = 150$ Wcm$^{-2}$). A piezo positioner raster scans the sample relative to the laser focus with dwell times per pixel below 1 s so as to minimize laser-induced changes in the sample. Emission is collected using the same focusing objective and is passed through a 405 nm dichroic filter (Semrock) and a 425 nm long-pass filter (Chroma). The emission is then collimated, split into two channels, and focused through two separate band-pass filters (Chroma: 730±25 nm; 560±20 nm) onto identical single-photon counting avalanche photodiodes (APD; Perkin Elmer).

Figure 2.7 shows an emission spectrum obtained from a ~2 µm wide strip of a single GO sheet. Two characteristic peaks are observed, located in the blue (~490 nm) and red (~690 nm) portions of the visible spectrum. This agrees with other reported ensemble GO spectra [27–29]. As described earlier, GO's broad emission is generally attributed to various causes including the presence of intrasheet $sp^2$ domain size distributions [24, 27], quasi-molecular ligand/$sp^2$ states [29], and heterogeneous carrier relaxation kinetics [28]. Associated single-GO sheet emission QYs—determined from correlated absorption/emission measurements—are on the order of ~1 % [21].

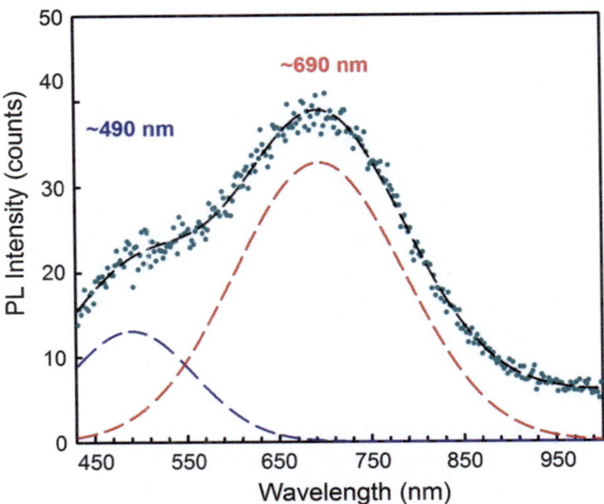


**Fig. 2.7** Emission spectrum of a single GO sheet (*solid green circles*) showing both *blue* and *red* components. The trace is fit to a sum of Gaussians (*dashed black line*) where two individual components (*dashed blue* and *red lines*) are extracted with center wavelengths of ~490 and ~690 nm. Traces offset for clarity. Reprinted with permission from McDonald, M.P., et al., *Nano Letters*, 2013, *13*, 5777–5784. Copyright 2013 American Chemical Society

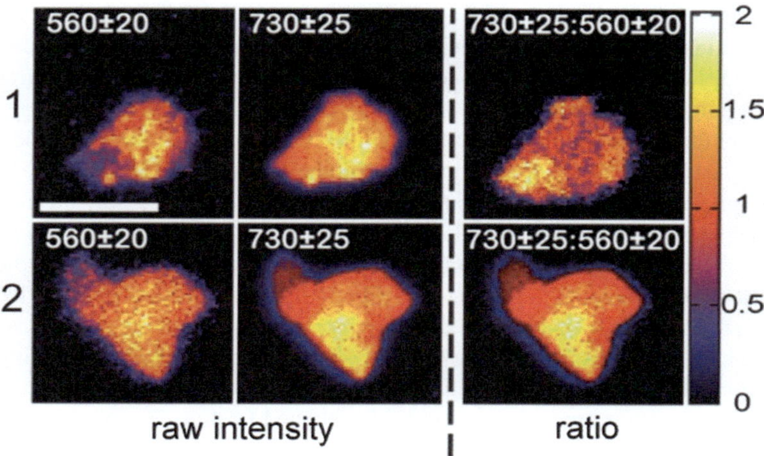



**Fig. 2.8** Confocal emission images of two different (*rows 1* and *2*) GO sheets under 405 nm excitation. In either case, the collected emission is split into *red* (730±25 nm) and *green* (560±20 nm) channels using band-pass filters (corresponding raw intensity images are to the *left* of the *dashed line*). The *rightmost* column shows ratio images of the two colors (730±25:560±20). Scale bar: 5 μm. Reprinted with permission from McDonald, M.P., et al., *Nano Letters*, 2013, *13*, 5777–5784. Copyright 2013 American Chemical Society

As with earlier single-sheet absorption images, single-sheet confocal emission maps exhibit substantial spatial heterogeneities. Figure 2.8 illustrates this with two sets of representative confocal images obtained in the green (560±20 nm) and red (730±25 nm) portions of the GO spectrum. Intensity variations exist in each image indicating that different parts of the same sheet emit different amounts of the same color. To illustrate, the lower left portion in Fig. 2.8(1) is much less emissive than the rest of the sheet. This could stem from local emission QY variations, spectral differences, or associated absorption heterogeneities. Correlated absorption/emission images, however, show no apparent link between absorption and emission magnitudes. Consequently, absorption heterogeneities (e.g., Fig. 2.5) alone do not account for the observed intensity variations in Fig. 2.8 [21]. Local emission efficiencies can be determined using spatially resolved emission lifetime experiments. However, such studies have not been conducted yet and are reserved for future investigations.

At this point, to reveal the existence of any intrasheet emission spectral differences, ratio maps are created by dividing single-color, confocal images (right column, 730±25:560±20). Resulting maps indicate the existence of significant intrasheet spectral heterogeneities which can be seen as bright and dim domains in the representative images shown in Fig. 2.8. Namely, the GO sheet in Fig. 2.8(1) possesses a large (~1 μm$^2$) portion that emits primarily red light (bottom left of sheet). The rest of the sheet emits predominantly green light. Similar spectral heterogeneities exist in other GO sheets, with a second example shown in Fig. 2.8(2). We find no specific trends in the location of these predominantly red- or green-emitting regions, whether at the edges or the center of the sheets. Analogous spatial and spectral heterogeneities appear when using images acquired with other red and blue wavelengths [33].

The spatially heterogeneous emission spectrum of single-layer GO resembles analogous absorption spectral heterogeneities shown earlier in Fig. 2.6. This further demonstrates that significant disorder exists in the optical response of single-layer GO, stemming from the presence of $sp^2$ domain size distributions as well as associated chemical heterogeneities.

## 2.4  GO Photoreduction

Having established the intrinsic absorption and emission optical response of single-layer GO, we now investigate their evolution during reduction. In particular, we find that GO's photophysical properties change when individual sheets are exposed to prolonged periods of widefield 405 nm laser excitation [33, 21]. We observe what appear to be three distinct behaviors (called "Regions" in what follows) that characterize GO's optical evolution. This is illustrated in Fig. 2.9a where the top and bottom rows show correlated absorption and emission images of a single-layer GO sheet during continuous 405 nm irradiation ($I_{exc}$ ~220 Wcm$^{-2}$) [21]. Three pairs of images depict GO's absorption/emission prior to irradiation, after exposure for 55 s, and after exposure for 1,390 s of 405 nm irradiation.

From the images, GO's absorption and emission are seen to readily evolve. The most notable changes seen are the overall increase (decrease) of its absorption (emission). To better illustrate these changes, though, Fig. 2.9b shows the time-dependent evolution of GO's absorption/emission, extracted from a single point on the sheet (circled). Extending these measurements to all pixels subsequently reveals corresponding spatially resolved intrasheet absorption/emission differences which will be described in more detail below.

From Fig. 2.9, it is apparent that GO's initial response upon irradiation (denoted Region 1 in what follows) is a significant drop in emission intensity with a concomitant rise in absorption magnitude. Specifically, within the first 55 s of exposure to 405 nm light the emission intensity decreases by an order of magnitude with a corresponding quantum yield (QY) change from 0.65 to 0.075 % [21]. The total associated 405 nm photon dose ($\Phi$) yielding this change is $\Phi = 2.5 \times 10^{22}$ photons/cm$^2$

$[ \Phi = I_{exc} \times \left( \dfrac{\lambda}{hc} \right) \times t$ where $\lambda$ is the excitation wavelength, $h$ is Planck's constant, $c$

is the speed of light in a vacuum, and $t$ is the exposure time]. In parallel, Fig. 2.9c shows that this drop in emission intensity is accompanied by a simultaneous ~190 nm blueshift of the spectrum. A corresponding rise in absorption leads to a change in GO's 520 nm absorption efficiency from ~0.2 %A at the start of the experiment to ~1.3 %A.

Beyond ~100 s of exposure to 405 nm light ($\Phi = 4.5 \times 10^{22}$ photons/cm$^2$), Fig. 2.9b shows that this trend reverses. In what follows, this behavior is referred to as Region 2 of GO's optical evolution (denoted by the shaded blue region in Fig. 2.9b).

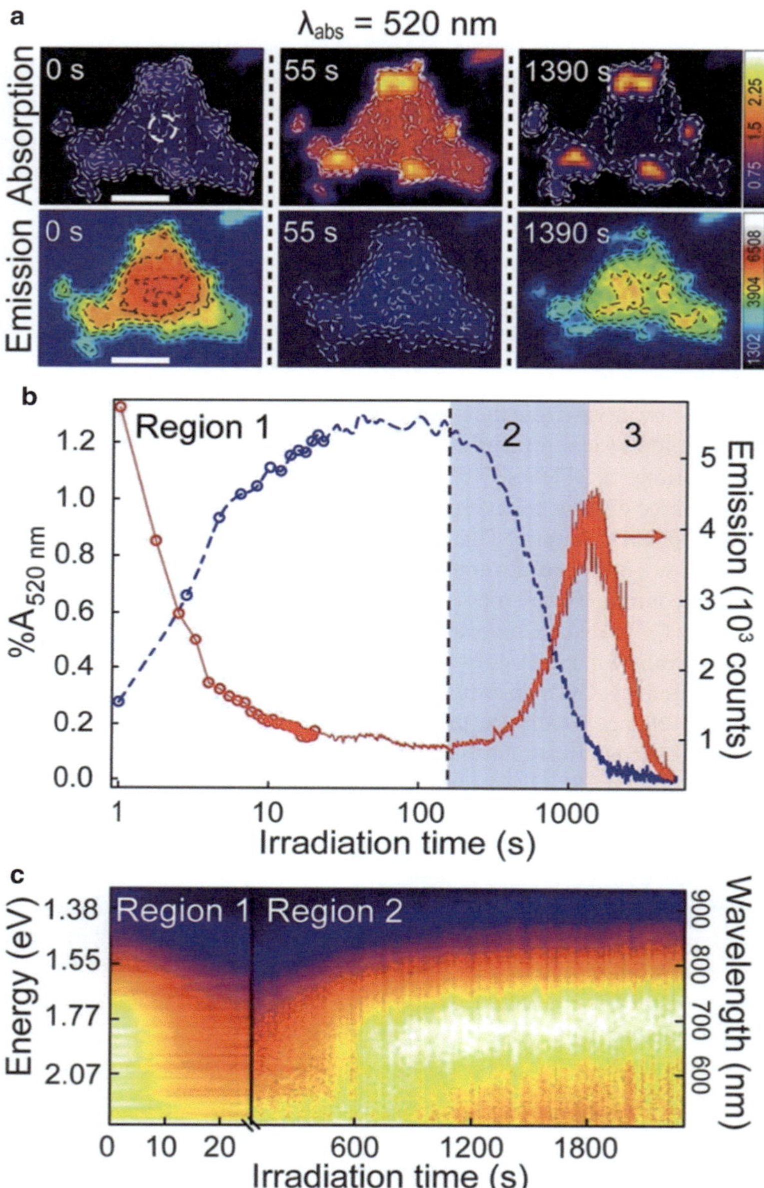



**Fig. 2.9** (**a**) Correlated absorption/emission images acquired during single-layer GO reduction ($\lambda_{abs}$ = 520 nm). Scale bar: 3 μm. (**b**) Absorption (*dashed red line/circles*) and emission (*solid blue line/circles*) intensities as a function of exposure time for the sheet featured in (**a**) where *shaded/colored portions* reflect *Regions 1, 2,* and *3*. (**c**). Waterfall plot of a sheet's normalized emission spectrum throughout *Regions 1* and *2*. Reprinted with permission from Sokolov, D.A., et al., *Nano Letters* 2014, *14*, 3172–3179. Copyright 2014 American Chemical Society

Namely, we observe that the sheet's emission brightens with emission QYs increasing from ~0.075 to 14 % [21]. A corresponding spectral redshift to ~725 nm (Fig. 2.9c) also occurs. At the same time, the sheet's absorption decreases by an order of magnitude, falling from a peak of ~1.3 to ~0.17 %.

Beyond this, there exists a third region of GO's optical response (called Region 3 and denoted by the shaded pink region in Fig. 2.9b) where both the emission and absorption decrease together. In the emission, we observe clear episodes of emission intermittency (i.e., intensity fluctuations that are not shot noise or laser intensity related) before irreversible photobleaching. In the absorption, the absorption efficiency continues to drop and eventually falls below the instrument's detection limit (~0.03 %A).

These correlated absorption/emission images have been acquired as follows. In brief, an absorption image is first taken and is then immediately followed by a 0.5–2-s period of widefield 405 nm laser exposure ($I_{exc}$ ~220 W cm$^{-2}$). Emission images are obtained during this time with an EMCCD camera. This absorption/emission imaging sequence is then repeated until no additional changes are seen in either the sheet's absorption or emission. This is referred to as the "stepwise" reduction method in the following discussions. Stop motion absorption/emission movies are subsequently created using these sequential images and can be found in ref. [21]. In this way, the spatial evolution of GO's optical properties has been mapped out throughout its optical life cycle.

An initial hypothesis for these observations is that they are linked to light-induced GO reduction. Supporting this, Fig. 2.9 shows that at the end of Region 1 the specimen's emission is weak and is blueshifted compared to that of pristine GO [28, 31]. Furthermore, the sample's absorption magnitude approaches that of graphene ~2.3 % [2]. This then suggests that prolonged 405 nm irradiation reduces GO and yields rGO at the end of Region 1. Continued light-induced chemistry then occurs in Region 2, causing rGO's emission to increase and its absorption to decrease. Loss of *both* the absorption and emission in Region 3 suggests that rGO is eventually destroyed.

To solidify this hypothesis and, in particular, the claim that Region 1 represents the photoreduction of GO to rGO, we probe the chemical nature of this transformation at the ensemble level using ensemble absorption spectroscopy, chemically specific X-ray photoelectron spectroscopy (XPS), Fourier transform infrared (FTIR) spectroscopy, and bulk electrical transport measurements.

To carry out these studies, an ensemble sample was prepared by irradiating a high-concentration, aqueous GO suspension with 405 nm light ($I_{exc}$ ~0.2 Wcm$^{-2}$) for 16.5 h. The associated photon dose is $\Phi$ ~2 × 10$^{22}$ photons/cm$^2$ and is equivalent to ~30 s of 405 nm laser exposure in the single-sheet experiment. This places the ensemble sample close to the end of Region 1 where we have claimed that rGO exists based on the single-sheet results in Fig. 2.9 [33].

Figure 2.10a first shows results of the ensemble absorption experiment where we see that GO's absorption gradually increases over the course of irradiation. The inset in Fig. 2.10a better illustrates this, revealing a near-linear rise in absorption at 405 nm. This change is accompanied by a noticeable decrease of the

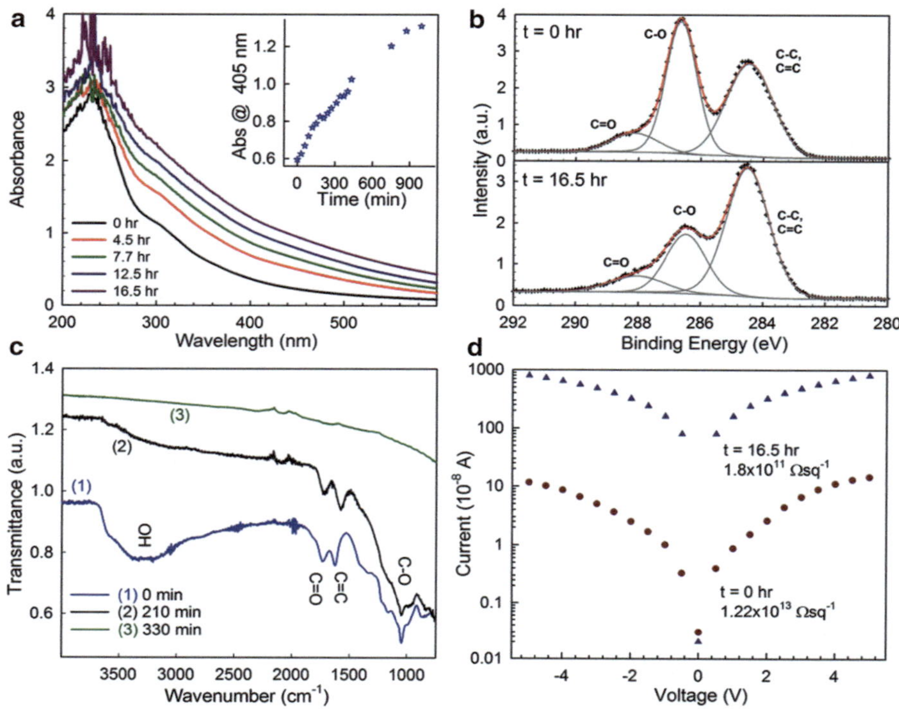



**Fig. 2.10** Ensemble GO/rGO measurements. (**a**) Absorption spectra of an aqueous suspension of GO throughout exposure to 16.5 h (990 min) of 0.2 Wcm$^{-2}$ 405 nm light (*inset*: absorption values at 405 nm as a function of irradiation time). (**b**) XPS C1s spectra of the GO ensemble before (*top*) and after (*bottom*) exposure to 16.5 h of 0.2 Wcm$^{-2}$ 405 nm light. (**c**) FTIR spectra of a GO film (1) exposed to 220 min (2) and 330 min (3) of 2 Wcm$^{-2}$ 405 nm laser irradiation. (**d**) Transport measurements of a GO ensemble before (*red circles*) and after (*blue triangles*) exposure to 16.5 h of 0.2 Wcm$^{-2}$ 405 nm light

$n-\pi^*$ resonance at ~320 nm, which is an observation consistent with earlier literature studies [23].

Next, Fig. 2.10b shows corresponding XPS spectra of the same ensemble. Chemical specificity in the measurement and specifically the ability to determine a sample's oxygen content come from the binding energy shift experienced by carbon 1 s (C1s) electrons in the presence of bound oxygen [44]. In our study, probing the C1s photoelectron spectrum reveals that peaks associated with oxidized carbon [44] substantially diminish after 16.5 h of exposure to 405 nm irradiation. Extracted estimates of the carbon-carbon/carbon-oxygen bond ratios (C:O) before and after 405 nm irradiation are 0.85 and 1.7, respectively. This indicates loss of oxygen during irradiation and supports the light-induced reduction of GO.

Additional, chemically specific evidence for reduction comes from associated FTIR spectra, obtained at given intervals during 405 nm irradiation ($I_{exc}$ ~2 Wcm$^{-2}$).

In this case, a dried GO film was used instead of the suspension to mitigate background OH vibrations associated with water. FTIR spectra in Fig. 2.10c show that IR transitions associated with GO's oxygen-containing functionalities all but disappear after 5.5 h of irradiation ($\Phi$ ~$6 \times 10^{22}$ photon/cm$^2$) [33]. These FTIR results clearly show that OH species are the first to be removed and are consistent with their basal plane abundance as well as with prior literature results [45, 46].

Finally, associated two-probe transport measurements of thin films made using these GO ensembles (Fig. 2.10d) show a sharp decrease in sheet resistance as a function of 405 nm irradiation. Specifically, experimental in-plane resistances drop nearly two orders of magnitude, falling from ~$1.2 \times 10^{13}$ $\Omega$sq$^{-1}$ to ~$1.8 \times 10^{11}$ $\Omega$sq$^{-1}$. This also agrees with prior ensemble GO reduction studies [47].

All of these results thus indicate that the optical evolution observed in Region 1 stems from photoreduction of GO to rGO. At the single-sheet level, this is evidenced by a marked increase in absorption magnitude with a simultaneous quenching and blueshifting of GO's emission.

## 2.5 Single-rGO Sheet Absorption Microscopy/Spectroscopy

Having established that photoreduction of GO to rGO occurs in Region 1, we now use single-sheet absorption and emission measurements to characterize rGO. Two experimental approaches are used to produce rGO sheets in the following analysis: "stepwise" and "continuous" reduction methods. The stepwise method is outlined in Sect. 2.4 above. Briefly, the 405 nm reduction laser is interrupted every 0.5–2 s for an absorption measurement to be carried out. Continuous reduction, by contrast, entails exposing the GO sheet to continuous 405 nm laser light for 50–60 s before acquiring an absorption image [21]. The sheet is therefore at the end of Region 1 before reduction is interrupted.

Both methods result in identical optical behaviors throughout reduction. However, we find that continuous reduction always produces slightly more absorptive rGO sheets [21]. As an example, resulting continuous reduction rGO sheets possess a maximum absorption of ~1.7 %A at 520 nm. Stepwise reduction absorption values, by contrast, only reach ~1.3 %A at 520 nm [21]. An identical trend is seen for all wavelengths where continuous reduction results in more absorptive rGO sheets [21]. This suggests that reoxidation potentially occurs during stepwise reduction, lowering the final absorption magnitude of resulting rGO sheets [21].

Figure 2.11 shows representative single-layer rGO absorption images acquired at (a) 405 nm, (b) 520 nm, and (c) 640 nm on a sheet produced using the continuous reduction method. The sheet's %A-values are 2.1 % (405 nm), 1.7 % (520 nm), and 1.4 % (640 nm). Associated $\alpha$-values are $\alpha_{405} = 2.1 \times 10^5$ cm$^{-1}$, $\alpha_{520} = 1.7 \times 10^5$ cm$^{-1}$, and $\alpha_{640} = 1.4 \times 10^5$ cm$^{-1}$. These numbers approach the graphene limit of 2.3 %A with a corresponding visible wavelength absorption coefficient of $\alpha = 5.8 \times 10^5$ cm$^{-1}$.

Similar to GO, sizable intrasheet absorption efficiency variations exist in rGO. In particular, Fig. 2.11c shows that a portion in the left center of the specific rGO sheet probed absorbs more 640 nm light (~1.5 %A) than the lower half of the sheet (~1.4 %A). Figure 2.11a–c also shows corresponding single-sheet α-value distributions as green and blue histograms. They clearly contribute to ensemble rGO α-value distributions (Fig. 2.11a–c; top gray histograms, average of 12 rGO sheets) and illustrate inhomogeneous broadening effects due to sheet-to-sheet variations in their extent of oxidation.

Figure 2.11d shows the absorption spectrum of a stepwise-reduced rGO sheet (green triangles). Superimposed are average rGO %A-values that include values from

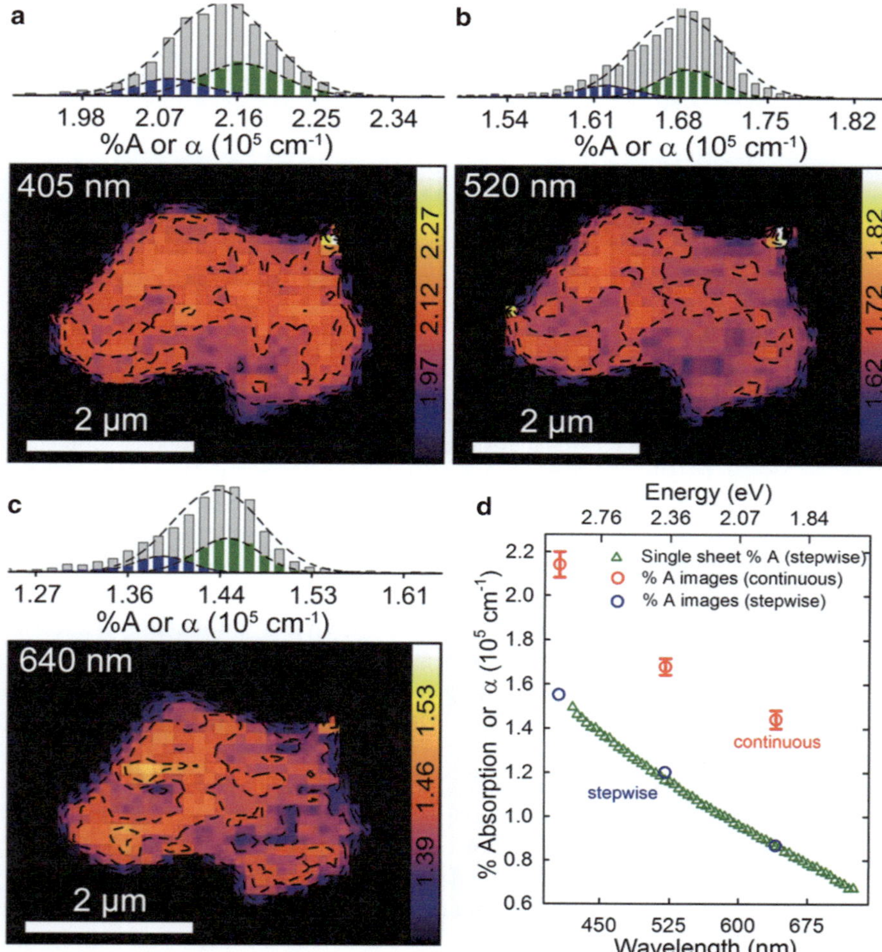

**Fig. 2.11** Single-rGO sheet %A images for (**a**) 405 nm, (**b**) 520 nm, and (**c**) 640 nm light. rGO was produced for (**a–c**) using the continuous reduction method. (**d**) Absorption spectrum of an rGO sheet reduced using the stepwise method (*green triangles*) compared to %A-values obtained from images of rGO sheets prepared with stepwise (*blue circles*) and continuous (*red circles*) reduction methods

both continuous (red circles) and stepwise (blue circles) reduction methods. In each case, average %A-values are obtained by averaging 12 individual sheet absorption images. At this point, given our ability to acquire full, single-sheet rGO spectra, a fit to the single-sheet absorption data (green triangles) yields the following estimate of rGO's wavelength-dependent absorption coefficients between $\lambda = 450$–700 nm:

$$\alpha\left(cm^{-1}\right) = 4.42 \times 10^5 - \lambda\left(1{,}160\right) + \lambda^2\left(1.39\right) - \lambda^3\left(7 \times 10^{-4}\right). \tag{2.3}$$

## 2.6  GO Photoreduction Mechanism

We now address how GO reduces upon prolonged exposure to 405 nm light. In this regard, GO photoreduction is well documented in the literature. It is often attributed to a photothermal mechanism [17, 18, 48–50] wherein the absorption of light results in the eventual heating of the sample. This follows from the conversion of excess absorbed energy into thermal energy through non-radiative relaxation. What results are sizable temperature jumps, especially when there exists poor heat dissipation into the environment. Local heating then provides the necessary energy required to initiate thermal deoxygenation reactions in GO [48].

Supporting this, temperatures as high as 500 °C have been reported in GO samples exposed to high-power flashlamps [17]. Similarly, pulsed lasers in the ultraviolet [51–53], visible [51], and NIR [54, 55] have all been used to reduce GO. Their short pulses and high peak powers result in estimated sample temperatures exceeding 1,000 °C [53, 54]. Thus, the combination of rapid energy uptake and poor thermal transfer into the local environment leads to the photothermal reduction of GO.

In our study, though, we use relatively low-intensity ($I_{exc}$ ~220–380 Wcm$^{-2}$) CW laser light to achieve GO reduction. A theoretical analysis—essentially an energy balance model based on known thermal conductance values and absorption strengths—provides a useful way to estimate GO's corresponding temperature rise ($\Delta T$) [33, 56]. What results is

$$\Delta T = \frac{I_{exc}\left(\dfrac{\%A_{405}}{100}\right)}{h_{sub}} \tag{2.4}$$

where $\%A_{405}$ is the %A at 405 nm and $h_{sub}$ is the interfacial thermal conductance between GO and SiO$_2$. Since GO $h_{sub}$ values are not available in the literature, we estimate it using values for graphene ($h_{sub} = 5 \times 10^7$ Wm$^{-2}$ K$^{-1}$) [57].

The model predicts $\Delta T$-values of <1 K for single-layer GO exposed to 380 Wcm$^{-2}$ of 405 nm CW excitation [33]. Consequently, resulting absolute temperatures are far below the ~230 °C temperatures needed to induce GO's thermal reduction [52]. Clearly, a photothermal mechanism alone does not account for the reduction seen in Fig. 2.9.

### 2.6.1 GO Photolysis

A handful of studies have implicated photolysis as an alternate, light-induced means of reducing GO [25, 49, 58]. In a photolytic mechanism, light is absorbed by a molecular resonance associated with an oxygen-containing functionality. The absorbed energy then dissipates through radiative or non-radiative channels or initiates a change in the chemical bond associated with the functionality. For GO, such changes include functional group migration and/or dissociation with theoretical activation energies ($E_a$) ranging from 0.28 to >1 eV [25, 58–61].

Specific GO functional group photolytic chemistries include direct OH, C–O–C, C=O, and COOH dissociation. Associated activation energies are 0.7 eV for OH [59], 1.9–2.1 eV for C–O–C [59, 62], and 3–4 eV for C=O and COOH groups [25, 58, 60]. Additional, more complicated, light-induced chemistries are also possible wherein multiple functional groups (primarily hydroxyls) interact to evolve $H_2O$ ($E_a$ ~0.28–0.46 eV) [61].

Functional group migration within GO's basal plane also occurs upon bond photolysis. In this case, hydroxyls (migration $E_a$ ~0.32 eV) and epoxides (migration $E_a$ ~0.9 eV) hop from carbon to carbon, leaving behind $sp^2$ hybridized carbons in their wake [61]. These groups eventually localize at defects/edges and subsequently dissociate through identical multiple functional group chemistries outlined above [25, 61].

All of these photolytic chemistries result in the evolution of $H_2O$, $O_2$, CO, and/or $CO_2$. This is consistent, with known products evolved during GO reduction [46]. Importantly, the release of water through either direct hydroxyl dissociation or through hydroxyl-hydroxyl interactions results in the reduction of GO without loss of carbon [61]. By contrast, other higher energy dissociation reactions, involving C–O–C/ C=O/COOH, all result in carbon abstraction through CO and/or $CO_2$ emission [58]. This induces the breakdown of GO/rGO's basal plane carbon network and will be important for explaining GO's optical response in Region 2 [49].

For photolysis to be a plausible mechanism to explain the reduction of GO, however, the employed 405 nm (3.06 eV) excitation must be on resonance with oxygen-containing functional group transitions. Supporting this, highly oxidized GO ensembles show $n–\pi^*$ transitions between 400 and 420 nm [25]. Additionally, molecular-like transitions extending out as far as 550 nm have been observed in GO photoluminescence excitation (PLE) spectra [29]. The existence of these latter transitions is supported by density functional theory (DFT) calculations, suggesting that oxygen-functionalized $sp^2$ clusters exhibit electronic transitions throughout the visible spectrum [27].

To conduct an initial check on the likelihood of a photolytic reduction mechanism, single-GO sheet excitation wavelength-dependent studies were conducted. Single-layer specimens were exposed to identical intensities of 405, 520, and 640 nm light ($I_{exc}$ ~700 $Wcm^{-2}$) with the idea that photolytic reduction should not occur under prolonged red light illumination. Results of this study support the proposed photolytic mechanism. Namely, in the case of 640 nm excitation, an initial emission decay characteristic of Region 1 is essentially absent. Emission intensities decrease by only ~10 % over the course of 1,800 s and no subsequent brightening is seen. These results suggest that the 640 nm excitation lies off-resonance with any ligand-related transition and agrees with prior ensemble absorption and PLE measurements [25, 29].

In stark contrast, both emission quenching and brightening are observed when individual GO sheets are exposed to 405 and 520 nm light. This is expected because both frequencies fall within the range of GO's oxygen-containing ligand resonances (400–550 nm) [25, 27, 29]. Furthermore, in the latter 520 nm case, this process occurs with kinetics that are ~5× slower, supporting the smaller abundance of these redder transitions [33]. Single-GO/rGO wavelength-dependent studies therefore qualitatively support a photolytic reduction mechanism.

### 2.6.2 Kinetics/Energetics of Photolysis

Given that GO's time-dependent absorption and emission trajectories (e.g., Fig. 2.9) are inherently linked to its photolytic reduction kinetics, photolytic rate constants can be extracted by analyzing absorption growth/emission decay traces. In Region 1, absorption traces (Fig. 2.9b) are empirically best fit with biexponential functions. A fast component contributes ~30 % to the total decay and has associated rate constants on the order of ~0.5 $s^{-1}$. A slow component also exists and accounts for ~70 % of the total decay. Associated rate constants are ~0.02 $s^{-1}$. To simplify subsequent analyses, rate constants from both contributions are combined into a single, weighted-average rate constant ($k_a^1$) [33, 21]. Region 1 emission decays are likewise analyzed with biexponential kinetics and yield weighted-average decay rate constants ($k_e^1$). Notably, $k_a^1$ and $k_e^1$ magnitudes are very similar, suggesting that both the absorption and emission evolve due to same physical process.

In Region 2, absorption decays are best described by single exponential functions. Consequently, a single $k$-value is extracted for the absorption ($k_a^2$). Emission growth trajectories are likewise fit with single exponential functions to give emission rate constants, $k_e^2$. Extracted $k_a^2$ and $k_e^2$ values are on the order of ~$2 \times 10^{-3}$ $s^{-1}$ and $3 \times 10^{-3}$ $s^{-1}$, respectively. Again, their magnitudes are similar and suggest that both absorption and emission changes in Region 2 arise from the same physical process. Furthermore, the magnitudes of Region 1 and Region 2 absorption/emission rate constants differ by 1–2 orders of magnitude, strongly suggesting that different chemistries lead to the diverging absorption/emission trends observed in Regions 1 and 2 of Fig. 2.9b [33, 21].

At this point, to link extracted $k_a^1$ and $k_e^1$ as well as $k_a^2$ and $k_e^2$ values to activation energies, temperature ($T$)-dependent photoreduction studies were conducted on individual GO sheets [33]. Sample temperatures ranging from 298 to 323 K were used with temperatures maintained with a resistively heated chamber and temperature controller (Lakeshore).

Results from these studies are shown in Fig. 2.12 where sheet-averaged Region 1/Region 2 rate constants, extracted from the $T$-dependent trajectories of several

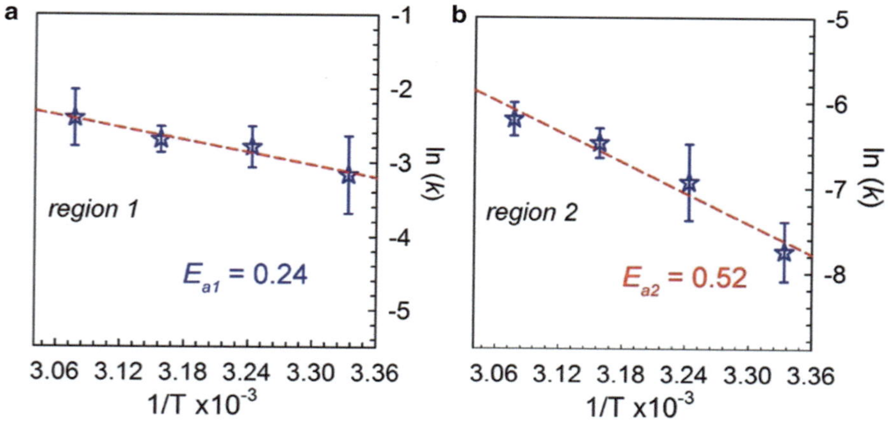


**Fig. 2.12** Arrhenius plots of GO reduction kinetics for temperatures between 298 and 323 K for (**a**) *Region 1* and (**b**) *Region 2. Error bars* indicate the standard deviation in rate constant values for five individual GO sheets at a given *T*. Reprinted with permission from McDonald, M.P., et al., *Nano Letters*, 2013, *13*, 5777–5784. Copyright 2013 American Chemical Society

different GO specimens, have been plotted as functions of temperature. Associated activation energies have subsequently been estimated using the Arrhenius relationship, where linear fits of experimental $\ln(k)$ versus $1/T$ give Region 1/Region 2 $E_a$-values of $E_{a1}$ ~0.24 eV and $E_{a2}$ ~0.52 eV, respectively [33]. Upper and lower bounds to these energies are $E_{a1}$ ~0.13–0.39 eV and $E_{a2}$ ~0.33–0.71 eV [33].

## 2.6.3 Photolytic Reduction Mechanism

Extracted $E_{a1}$ and $E_{a2}$ values enable us now to construct a more refined physical picture for the photolytic chemistries occurring in Regions 1, 2, and 3. Namely, the relatively small activation energies found in Region 1 ($E_{a1}$ ~0.13–0.39 eV) suggest hydroxyl migration ($E_a$ ~0.32 eV) as the dominant process responsible for GO reduction [61]. In this regard, we suggest that photolysis-induced OH migration leads to eventual OH localization at GO defects and edges whereupon corresponding increases of the local OH concentration enable multiple-functional group OH interactions to occur. This ultimately leads to the evolution of $H_2O$ as well as the concomitant reduction of GO [61]. Such multiple-functional group reactions include [61]

$$2R - OH \xrightarrow{h\nu} R - O - R + H_2O(g) \tag{2.5}$$

$$3R - OH + R - O - R \xrightarrow{h\nu} R - C - O* \rightarrow 2R - C = O + R - OH + H_2O(g) \tag{2.6}$$

where R represents the GO basal plane. Associated activation energies are 0.45 eV (Eq. 2.5) and 0.48 eV (step 1) as well as 0.30 eV (step 2) for Eq. (2.6).

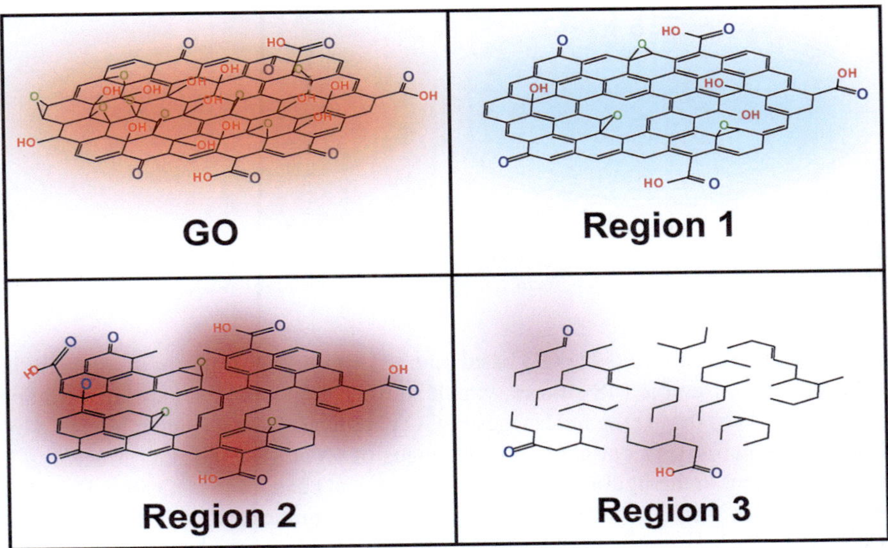


**Fig. 2.13** Scheme depicting GO/rGO during each phase of its photolytic reduction

Other possible multiple functional group reactions exist and are described in more detail in ref. [61].

Since no carbon is lost in this phase, resulting rGO sheets consist of highly interconnected $sp^2$ domains. From an optical standpoint, what results is a weak, blueshifted emission spectrum for rGO, since an interconnected $sp^2$ network favors fast non-radiative recombination following photoexcitation [28]. In tandem, larger single-sheet absorption efficiencies result. This rationalizes why experimental %A values at the end of Region 1 approach the graphene limit of 2.3 %A [2, 21]. Figure 2.13 summarizes this by showing a schematic depiction of an rGO sheet at the end of Region 1.

The association of OH chemistries to Region 1's photoreaction mechanism is supported by ensemble mass spectrometry (MS) measurements, which show strong $H_2O$ evolution during the initial stages of GO photolysis [46]. It is also supported by ensemble FTIR results shown in Fig. 2.10c where OH removal is clearly seen to occur during initial exposure to 405 nm radiation.

Region 2's larger $E_a$-values ($E_{a2}$ ~0.33–0.71 eV) suggest that other photolytic processes simultaneously coexist with Region 1's OH chemistries. Based on known photolytic reactions, these chemistries likely involve direct OH dissociation (0.7 eV), C–O–C migration (0.9 eV), or even C–O–C/C=O/COOH dissociation (>1 eV). Most of these latter processes involve carbon abstraction through the evolution of CO and/or $CO_2$ [46, 49, 58]. This is illustrated in the following reactions [58]:

$$R-CH=O \xrightarrow{hv} RH+CO \tag{2.7}$$

$$R-COH=O \xrightarrow{hv} RH+CO_2 \tag{2.8}$$

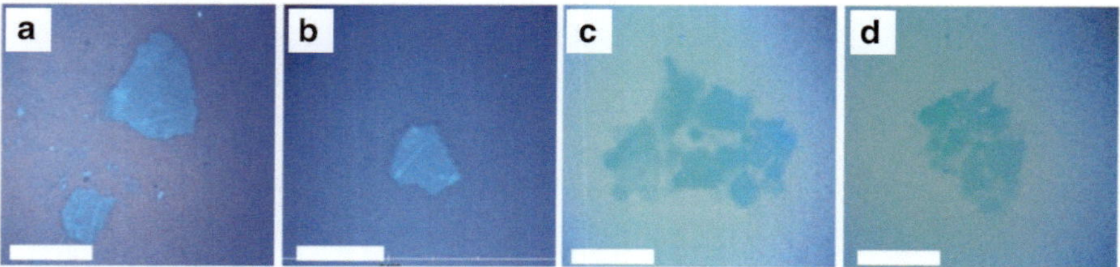

**Fig. 2.14** Optical contrast images of GO on Si/SiO$_2$ (300 nm) substrates before (**a**, **b**) and after (**c**, **d**) 405 nm laser irradiation (2.8 Wcm$^{-2}$; $\Phi = 2.1 \times 10^{23}$ photons/cm$^2$). Scale bars: 10 μm

where, again, R represents the GO basal plane. As with Region 1, our assignment is supported by ensemble MS studies, which show that CO and CO$_2$ are the primary products evolved during the final stages of GO photolysis [46]. It is also supported by Fig. 2.14 which shows optical contrast images of GO (a, b) and rGO in Region 2 (c, d). Apparent in the figures are clear morphological changes to rGO after extended 405 nm irradiation [33]. Holes and basal plane discontinuities are evident. This observation is also consistent with other AFM and TEM studies, which show the appearance of similar holes/defects in rGO following reduction [19, 49].

Optically, what results in Region 2 is a gradual loss of absorption due to carbon abstraction. rGO's emission, however, initially rises, since fragmentation of its aromatic network implies a loss of fast non-radiative recombination channels that induce emission quenching and blueshifts [28]. rGO's emission QY therefore increases and is accompanied by an apparent spectral redshift [28]. In essence, Region 2 unmasks the $sp^2$ domain size growth originally established in Region 1 with resulting near-isolated $sp^2$/ligand clusters behaving as disparate, quasi-molecular entities. Corroborating this, fluorescence intermittency (reminiscent of quantum dot and single-molecule "blinking") is apparent in Region 2 (Fig. 2.9b) [33].

The destructive nature of Region 2 photolytic chemistries also means that this state of rGO is only temporary. Continued photolysis results in additional carbon abstraction. This leads to continued absorption losses and, in the emission, ultimately results in irreversible photobleaching. The eventual photolytic destruction of rGO thus completes GO's optical life cycle and rationalizes its optical response in Region 3 of Fig. 2.9. Figure 2.13 summarizes this by providing a schematic description of rGO in Regions 2 and 3.

## 2.7 Temporal/Spatial Evolution of Photolysis

Having associated observed changes in GO's absorption/emission with photoreduction and having posited likely photolytic chemistries responsible for these changes, we now rationalize how GO's reduction spreads across its basal plane. This is done by analyzing sequential GO absorption and emission images acquired during photolysis. As seen earlier (e.g., Figs. 2.5, 2.6, and 2.8), significant optical heterogeneity exists within individual GO sheets. We have attributed this to the existence of chemically segregated regions on the ~$\mu m^2$ scale within GO's basal plane. What results are spatially heterogeneous reduction chemistries that can be observed in real time using single-sheet absorption and emission imaging. Corresponding analyses of spatially resolved emission quenching and absorption enhancements then lead to local estimates of GO reduction kinetics. In turn, these local rate constant variations allow us to rationalize the spatial evolution of GO's reduction, which empirically appears to "flow" across individual sheets after having nucleated at random reduction "hot spots."

In practice, spatially resolved, kinetic rate constant maps are constructed by analyzing trajectories associated with every pixel (a 250 nm×250 nm region) in an absorption/emission movie. Figure 2.9b shows one such representative trajectory. Region 1 and 2 absorption/emission fits are carried out using home-written software (a Java applet or a Visual Basic macro; source code available in refs. [33] and [21]). Region 3 analyses are omitted since we have previously attributed the kinetics here to the destruction of rGO.

Figure 2.15 shows resulting absorption (a, c) and emission (b, d) rate constant maps for a single GO sheet undergoing photoreduction (the same sheet featured in Fig. 2.9a). Region 1 absorption/emission maps are shown in the left column. Corresponding Region 2 absorption/emission maps are shown in the right column. In all cases, clear intrasheet $k_a/k_e$ variations are apparent. For example, the absorption growth ($k_a^1$) rate map in Fig. 2.15a shows a large area in the upper right-hand corner of the sheet with $k_c^1 \sim 0.16 s^{-1}$. A second section in the bottom left of the sheet shows $k_a^1 \sim 0.06 s^{-1}$. Similar heterogeneities are present in other maps.

As expected from our discussion in Section 2.6.3, the analysis reveals spatial correlations between $k_a$ and $k_e$ rate constants within a given Region. To illustrate, in Region 1 a high-$k$ ridge exists in the upper right-hand portion of both Fig. 2.15a $\left( k_a^1 \sim 0.16 s^{-1} \right)$ and Fig. 2.15b $\left( k_e^1 \sim 0.07 s^{-1} \right)$. Other correlations between $k_a^1$ and $k_e^1$ are evident in both the middle and bottom right sections of the same sheet. In Region 2, analogous absorption/emission $k$-value correlations exist. For example, the bottom right protrusion in Fig. 2.15c, d exhibits similarly peaked $k_a^2$ and $k_e^2$ values.

Furthermore, as seen earlier, extracted absorption/emission rate constants for a given region take the same order of magnitude. Specifically, average Region 1 absorption/emission rate constants in Fig. 2.15 are $\left\langle k_a^1 \right\rangle = 0.10 s^{-1}$ and $\left\langle k_e^1 \right\rangle = 0.04 s^{-1}$. In Region 2 they are $\left\langle k_a^2 \right\rangle = 1.75 \times 10^{-3} s^{-1}$ and $\left\langle k_e^2 \right\rangle = 3.8 \times 10^{-3} s^{-1}$. The good agreement between these $k_a$ and $k_e$ pairs then suggests that, within a given region, changes in the absorption and emission arise from the same physical process. Differing magnitudes of Region 1/Region 2 $k_a/k_e$ values simultaneously suggest different photolytic chemistries being responsible for the behavior in these regions.

Proposed Region 1 and Region 2 photolytic chemistries have been outlined in Section 2.6.3 by linking rate constants to activation energies using temperature-dependent measurements. Within the context of the spatially resolved measurements described here, this leads to analogous $E_a$ maps constructed from $k_a/k_e$ maps shown in Fig. 2.15. At this point, the most important thing to note is that in Region 1 of Fig. 2.15, peaks in the rate constant map are associated with the smallest activation

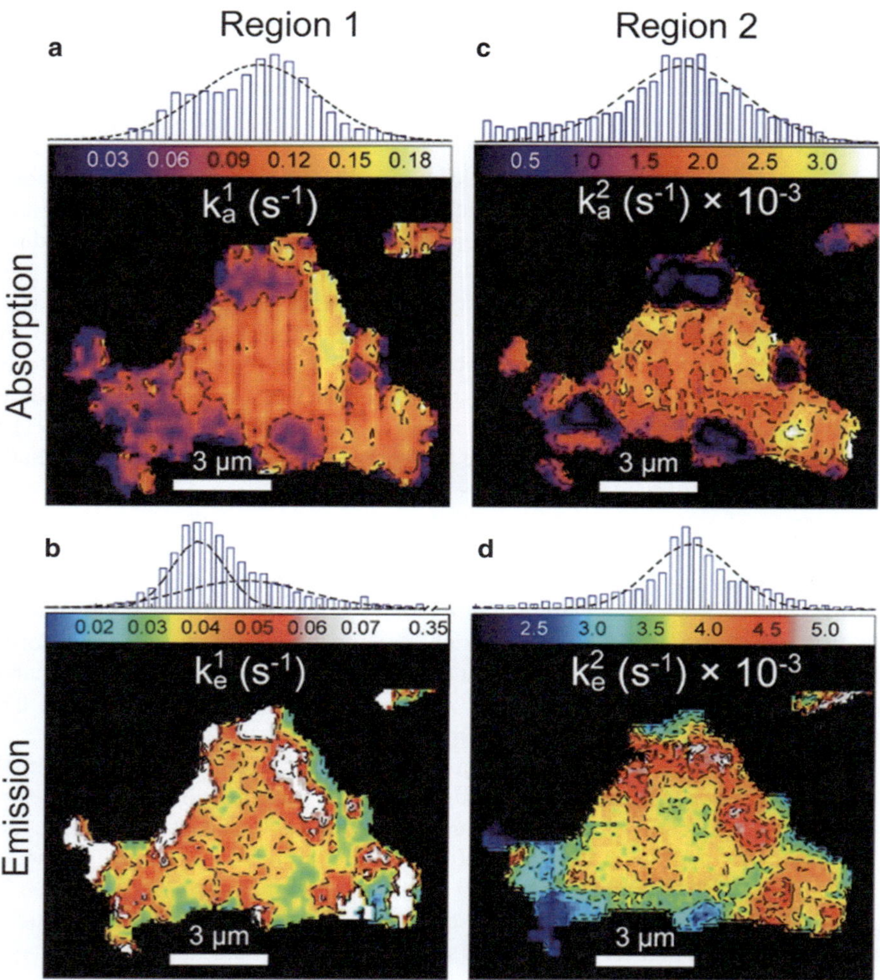



**Fig. 2.15** Absorption (at 520 nm) and emission rate constant maps obtained for (**a**, **b**) *Region 1* and (**c**, **d**) *Region 2*. *Histograms* illustrate absorption/emission rate constant distributions present within the GO sheet. Reprinted with permission from Sokolov, D.A., et al., *Nano Letters* 2014, *14*, 3172–3179. Copyright 2014 American Chemical Society

energies ($E_a$ ~0.20 eV). It is these very regions which "nucleate" reduction in GO photolytic absorption/emission movies.

Given that we have previously attributed GO reduction in Region 1 to OH photolytic chemistries and specifically to OH migration and multiple OH group reactions (Eqs. 2.5 and 2.6) [61], this suggests that GO reduction commences in regions of the basal plane possessing relatively large OH concentrations. Once nucleated, reduction "spreads" outwards since photolytic reduction in neighboring regions, with fewer

OH groups, involves OH migration to local defects/edges followed by multiple OH group chemistries, which ultimately lead to the evolution of water. From a kinetic standpoint, what results is an effective activation energy/rate constant that is larger/ smaller than at nucleation sites due to the fact that multiple processes are involved in GO reduction in these locations. This suggests that the spatially heterogeneous rate constant/activation energy maps seen in Fig. 2.15 reflect the large-scale distribution of OH functional groups within GO's basal plane.

Some evidence for this hypothesis exists. Namely, the absorption growth/ emission decay traces in Region 1 are best fit to biexponential functions. The non-single exponential nature of these traces thus implies that not one but rather several photolytic processes underlie the observed kinetics. Thus, there exists a need for future experiments that can better unravel the various, unresolved, photolytic processes present in Region 1. Furthermore, a need exists for experiments that possess spatially resolved chemical specificity and which can better assess large-scale functional group distributions in GO/rGO.

At this point, an analysis of Region 2 kinetics is more complicated. However, similar "nucleation" and photolysis fronts are seen in Region 2 photoreduction movies. Our prior analysis of the kinetics here has suggested that carbon-abstracting photolytic chemistries are at play (Eqs. 2.7 and 2.8). Hence, the rGO produced in Region 1 continues to be reduced in Region 2 and, in tandem, undergoes gradual breakdown of its $sp^2$ network.

To explain the visual propagation of Region 2 photolytic chemistries, we first posit that these latter chemistries begin in aromatic $sp^2$ domain-rich regions. In principle, a correlation could exist between nucleation sites in Regions 1 and 2 provided that OH-rich regions produce large, strongly absorbing $sp^2$ domains when reduced. However, in practice, this need not be the case. At present, the data is ambiguous and hence spatial correlations between Region 1 and 2 nucleation sites cannot be established.

Next, in sections of individual rGO sheets containing strongly absorbing $sp^2$ domains, associated photolytic processes propagate due to temperature gradients that exist between these domains and neighboring $sp^3$ hybridized regions. In this regard, given that semi-isolated $sp^2$ clusters are more absorptive than their $sp^3$ surroundings and have modest emission quantum yields, excitation of these clusters leads to local heating. What results are $\Delta T$ gradients of up to ~4 K based on the energy balance model discussed earlier (Eq. 2.4) [33]. While this temperature increase is not enough to directly induce thermal reduction, it can enhance the rate of an activated process through local variations of the chemical rate constant. As an illustration, rate constant increases of 20–45 % are possible when activation energies are in the range of $E_a$ ~0.52–70 eV [33]. Consequently, partially oxidized $sp^3$-rich domains in close proximity to strongly absorbing and "hot" $sp^2$ regions should undergo photolytic chemistries faster than more isolated $sp^3$ domains. Thus, with continued reduction leading to corresponding increases in $sp^2$ domain number as well as size, a $\Delta T$–$k$ feedback loop ensues, causing the photolysis front to spread and eventually consume the entire rGO sheet.

## 2.8 Conclusions

GO and rGO have been studied extensively using various structural and chemical microscopies. They have also been studied extensively using ensemble absorption and emission spectroscopies. Recently, a growing number of atomic resolution microscopies have probed the structural and electronic properties of these materials. However, to date, little work has been done to understand their photophysics at the microscopic level. A number of reasons exist for wanting to do this. Perhaps the most pressing involves a desire to develop a deeper understanding of GO's photoreduction mechanism.

In what has preceded, we have described work to study the optical response of individual single-layer GO and rGO using spatially resolved absorption and emission microscopies/spectroscopies. We find significant spatial and spectral heterogeneities within the optical response of individual GO and rGO sheets. The origin of these absorption and emission heterogeneities likely stems from the presence of varying-sized $sp^2$ domains as well as chemical disorder within GO's basal plane. The single-sheet microscopies/spectroscopies we have developed are quantitative. Consequently, we have established the first experimental accounting of GO and rGO absorption parameters across the visible.

More intriguingly, under prolonged illumination with 405 nm light, dynamics in the optical response of single-layer GO are seen. Namely, an apparent emission quenching and spectral blueshifting are observed. In tandem, a noticeable increase of the absorption occurs. Afterwards, the absorption decreases while the emission increases markedly and is accompanied by a sizable redshift of the spectrum. Finally, the end of GO's optical life cycle is characterized by a continued loss of absorption along with apparent emission intermittency prior to irreversible photobleaching.

By linking various ensemble and single-sheet experiments, we have established that GO reduction to rGO is characterized by aforementioned emission quenching and absorption growth, seen in the initial stages of GO's optical life cycle. The underlying mechanism responsible for this photoreduction is thought to be photolytic in nature—not photothermal. Specifically, energy balance estimates show that only modest temperature jumps are achieved in GO following the absorption of light. Detailed kinetics studies on GO's spatially resolved absorption and emission dynamics subsequently establish rate constants and associated activation energies consistent with known OH photolytic chemistries. Consequently, we attribute observed spatially heterogeneous GO reduction chemistries to photolytic OH processes and, by the same token, to large-scale OH concentration variations across GO's basal plane.

Prolonged illumination leads to the eventual destruction of rGO. This occurs through additional, carbon-abstracting photolytic chemistries that follow GO's photoreduction. Initially, these processes result in rGO's apparent emission brightening. Their destructive nature, though, appears through commensurate absorption losses. Beyond a certain point, continued photolysis leads to the physical destruction of rGO as observed optically through complete loss of its absorption along with irreversible photobleaching. This is also seen structurally through unambiguous morphological changes in optical images of rGO after prolonged illumination and in corresponding AFM and TEM measurements. The role of these latter photolytic processes has again been established using detailed, spatially resolved kinetic studies on rGO wherein we find observed rate constants and associated activation energies consistent with those of known carbon-abstracting photolytic chemistries.

We have therefore constructed a detailed, microscopic picture of GO's optical properties. However, there are still a number of unresolved questions within the field. For example, the origin of GO's emission is still debated. Evidence in the literature suggests that it can arise from direct recombination in $sp^2$ clusters, quasi-molecular ligand/$sp^2$ states, or other photophysical processes. Unfortunately, the optical approach illustrated in this chapter is limited to a ~200 nm resolution, and is thus unable to interrogate individual $sp^2$ domains. An intriguing possibility for overcoming this limitation could be realized through single-GO/rGO sheet near-field scanning optical microscopy.

The measurements outlined in this chapter have also established unprecedented insight into GO's photoreduction mechanism. In turn, these results provide the foundation for future studies aiming to better understand light-initiated GO-to-rGO interconversion. Beyond this, these studies and others like them may reveal opportunities for better controlling GO's reduction, ultimately enabling spatially selective patterning and hence local control of GO's optical and electrical properties. Future spectroscopic studies that elucidate the reversibility and fidelity of these photochemical processes are thus crucial to next-generation GO/rGO device fabrication. Finally, the developed absorption/emission techniques discussed here open up opportunities for studying the optical properties of other two-dimensional systems in ways free of constraints limiting existing, emission-based single-particle microscopies.

**Acknowledgements** MK thanks the ACS PRF (Type ND 51675) and the Army Research Office (W911NF-12-1-0578) for financial support. JHH thanks CONICET for the international cooperation funds [D979 (25-03-2013)], FONCyT for research grant P.BID2012 PICT-2041, and University of Buenos Aires for grants UBACYT 2015-2017 20020130100643BA and UBACYT 01-w971.

# References

1. Novoselov KS, Geim AK, Morozov SV et al (2004) Electric field effect in atomically thin carbon films. Science 306:666–669
2. Nair RR, Blake P, Grigorenko AN et al (2008) Fine structure constant defines visual transparency of graphene. Science 320:1308
3. Hwang EH, Adam S, Das Sarma S (2007) Carrier transport in two-dimensional graphene layers. Phys Rev Lett 98:186806-1–186806-4
4. Wu Z-S, Ren W, Gao L et al (2009) Synthesis of graphene sheets with high electrical conductivity and good thermal stability by hydrogen arc discharge exfoliation. ACS Nano 3:411–417

5. Balandin AA, Ghosh S, Bao W et al (2008) Superior thermal conductivity of single-layer graphene. Nano Lett 8:902–907

6. Kim KS, Zhao Y, Jang H et al (2009) Large-scale pattern growth of graphene films for stretchable transparent electrodes. Nature 457:706–710

7. Wang X, Ouyang Y, Li X et al (2008) Room-temperature all-semiconducting sub-10-nm graphene nanoribbon field-effect transistors. Phys Rev Lett 100:206803-1–206803-4

8. Chae SJ, Gunes F, Kim KK et al (2009) Synthesis of large-area graphene layers on poly-nickel substrate by chemical vapor deposition: wrinkle formation. Adv Mater 21:2328–2333

9. Emtsev KV, Bostwick A, Horn K et al (2009) Towards wafer-size graphene layers by atmospheric pressure graphitization of silicon carbide. Nat Mater 8:203–207

10. Zhang Y-L, Guo L, Xia H et al (2014) Photoreduction of graphene oxides: methods, properties, and applications. Adv Opt Mater 2:10–28

11. Brodie BC (1859) On the atomic weight of graphite. Philos Trans R Soc Lond 149:249–259

12. Hummers WS, Offeman RE (1958) Preparation of graphitic oxide. J Am Chem Soc 80: 1339–1339

13. Lerf A, He H, Forster M et al (1998) Structure of graphite oxide revisited. J Phys Chem B 102: 4477–4482

14. Krishnan D, Kim F, Luo J et al (2012) Energetic graphene oxide: challenges and opportunities. Nano Today 7:137–152

15. Stankovich S, Dikin DA, Piner RD et al (2007) Synthesis of graphene-based nanosheets via chemical reduction of exfoliated graphite oxide. Carbon 45:1558–1565

16. Wang X, Zhi L, Müllen K (2008) Transparent, conductive graphene electrodes for dye-sensitized solar cells. Nano Lett 8:323–327

17. Gilje S, Dubin S, Badakhshan A et al (2010) Photothermal deoxygenation of graphene oxide for patterning and distributed ignition applications. Adv Mater 22:419–423

18. Sokolov DA, Shepperd KR, Orlando TM (2010) Formation of graphene features from direct laser-induced reduction of graphite oxide. J Phys Chem Lett 1:2633–2636

19. Erickson K, Erni R, Lee Z et al (2010) Determination of the local chemical structure of graphene oxide and reduced graphene oxide. Adv Mater 22:4467–4472

20. Paredes JI, Villar-Rodil S, Solis-Fernandez P et al (2009) Atomic force and scanning tunneling microscopy imaging of graphene nanosheets derived from graphite oxide. Langmuir 25:5957–5968

21. Sokolov DA, Morozov YV, McDonald MP et al (2014) Direct observation of single layer graphene oxide reduction through spatially resolved, single sheet absorption/emission microscopy. Nano Lett 14:3172–3179

22. Zhou Y, Bao Q, Varghese B et al (2010) Microstructuring of graphene oxide nanosheets using direct laser writing. Adv Mater 22:67–71

23. Eda G, Lin Y-Y, Mattevi C et al (2010) Blue photoluminescence from chemically derived graphene oxide. Adv Mater 22:505–509

24. Shang J, Ma L, Li J et al (2012) The origin of fluorescence from graphene oxide. Sci Rep 2: 792-1–792-8

25. Andryushina NS, Stroyuk OL, Dudarenko GV et al (2013) Photopolymerization of acrylamide induced by colloidal graphene oxide. J Photochem Photobiol A 256:1–6

26. Li D, Muller MB, Gilje S et al (2008) Processable aqueous dispersions of graphene nanosheets. Nat Nanotechnol 3:101–105

27. Chien C-T, Li S-S, Lai W-J et al (2012) Tunable photoluminescence from graphene oxide. Angew Chem Int Ed 51:6662–6666

28. Exarhos AL, Turk ME, Kikkawa JM (2013) Ultrafast spectral migration of photoluminescence in graphene oxide. Nano Lett 13:344–349

29. Galande C, Mohite AD, Naumov AV et al (2011) Quasi-molecular fluorescence from graphene oxide. Sci Rep 1:85-1–85-5

30. Luo Z, Vora PM, Mele EJ et al (2009) Photoluminescence and band gap modulation in graphene oxide. Appl Phys Lett 94:111909-1–111909-3

31. Loh KP, Bao Q, Eda G et al (2010) Graphene oxide as a chemically tunable platform for optical applications. Nat Chem 2:1015–10024
32. Gokus T, Nair RR, Bonetti A et al (2009) Making graphene luminescent by oxygen plasma treatment. ACS Nano 3:3963–3968
33. McDonald MP, Eltom A, Vietmeyer F et al (2013) Direct observation of spatially heterogeneous single-layer graphene oxide reduction kinetics. Nano Lett 13:5777–5784
34. Cuong TV, Pham VH, Tran QT et al (2010) Photoluminescence and raman studies of graphene thin films prepared by reduction of graphene oxide. Mater Lett 64:399–401
35. Chen J-L, Yan X-P (2010) A dehydration and stabilizer-free approach to production of stable dispersion of graphene nanosheets. J Mater Chem 20:4328–4332
36. Hou X-L, Li J-L, Drew SC et al (2013) Tuning radical species in graphene oxide in aqueous solution by photoirradiation. J Phys Chem C 117:6788–6793
37. Li J-L, Kudin KN, McAllister MJ et al (2006) Oxygen-driven unzipping of graphitic materials. Phys Rev Lett 96:176101-1–176101-4
38. Giblin J, Vietmeyer F, McDonald MP et al (2011) Single nanowire extinction spectroscopy. Nano Lett 11:3307–3311
39. Vietmeyer F, McDonald MP, Kuno M (2012) Single nanowire microscopy and spectroscopy. J Phys Chem C 116:12379–12396
40. McDonald MP, Vietmeyer F, Kuno M (2012) Direct measurement of single CdSe nanowire extinction polarization anisotropies. J Phys Chem Lett 3:2215–2220
41. Vietmeyer F, Tchelidze T, Tsou V et al (2012) Electric field-induced emission enhancement and modulation in individual CdSe nanowires. ACS Nano 6:9133–9140
42. McDonald MP, Vietmeyer F, Aleksiuk D et al (2013) Supercontinuum spatial modulation spectroscopy: detection and noise limitations. Rev Sci Instrum 84:113104-1–113104-7
43. Geim AK, Novoselov KS (2007) The rise of graphene. Nat Mater 6:183–191
44. Dreyer DR, Park S, Bielawski CW et al (2010) The chemistry of graphene oxide. Chem Soc Rev 39:228–240
45. Chen W, Yan L (2010) Preparation of graphene by a low-temperature thermal reduction at atmosphere pressure. Nanoscale 2:559–563
46. Shulga YM, Martynenko VM, Muradyan VE et al (2010) Gaseous products of thermo- and photo-reduction of graphite oxide. Chem Phys Lett 498:287–291
47. Eda G, Fanchini G, Chhowalla M (2008) Large area ultrathin films of reduced graphene oxide as a transparent and flexible electronic material. Nat Nanotechnol 3:270–274
48. Cote LJ, Cruz-Silva R, Huang J (2009) Flash reduction and patterning of graphite oxide and its polymer composite. J Am Chem Soc 131:11027–11032
49. Matsumoto Y, Koinuma M, Ida S et al (2011) Photoreaction of graphene oxide nanosheets in water. J Phys Chem C 115:19280–19286
50. Guo H, Peng M, Zhu Z et al (2013) Preparation of reduced graphene oxide by infrared irradiation induced photothermal reduction. Nanoscale 5:9040–9048
51. Abdelsayed V, Moussa S, Hassan HM et al (2010) Photothermal deoxygenation of graphite oxide with laser excitation in solution and graphene-aided increase in water temperature. J Phys Chem Lett 1:2804–2809
52. Sokolov DA, Rouleau CM, Geohegan DB et al (2013) Excimer laser reduction and patterning of graphite oxide. Carbon 53:81–89
53. Huang L, Liu Y, Ji L-C et al (2011) Pulsed laser assisted reduction of graphene oxide. Carbon 49:2431–2436
54. Trusovas R, Ratautas K, Raciukaitis G et al (2013) Reduction of graphite oxide to graphene with laser irradiation. Carbon 52:574–582
55. Zhang Y, Guo L, Wei S et al (2010) Direct imprinting of microcircuits of graphene oxides film by femtosecond laser reduction. Nano Today 5:15–20
56. Wang D, Carlson MT, Richardson HH et al (2011) Absorption cross section and interfacial thermal conductance from an individual optically excited single-walled carbon nanotube. ACS Nano 5:7391–7396

57. Jeong H-K, Lee YP, Jin MH et al (2009) Thermal stability of graphite oxide. Chem Phys Lett 470:255–258
58. Plotnikov VG, Smirnov VA, Alfimov MV et al (2011) The graphite oxide photoreduction mechanism. High Energy Chem 45:411–415
59. Lahaye RJWE, Jeong HK, Park CY et al (2009) Density functional theory of graphite oxide for different oxidation levels. Phys Rev B 79:125435-1–125435-8
60. Smirnov VA, Shul'ga YM, Denisov NN et al (2012) Photoreduction of graphite oxide at different temperatures. Nanotechnol Russia 7:156–163
61. Ghaderi N, Paressi M (2010) First-principle study of hydroxyl functional groups on pristine, defected graphene, and graphene epoxide. J Phys Chem C 114:21625–21630
62. Sorescu DC, Jordan KD (2001) Theoretical study of oxygen adsorption on graphite and the (8,0) single-walled carbon nanotube. J Phys Chem B 105:11227–11232

# Chapter 3
# The Chemistry of Graphene Oxide

Wei Gao

**Abstract** In this chapter, we discuss a variety of chemical reactions introduced for GO. Among all studies on the chemistry of GO, the largest portion focused on the reduction of GO back to graphene, mainly due to its high relevance to graphene and the gold rush of graphene research over the last decade. However, doping, functionalization and cross-linking of GO are equally, if not more, interesting to chemists, since GO is a giant model compound of polycyclic aromatic hydrocarbon (PAH) oxides. Here, we start with a thorough comparison between various reducing recipes for GO, and follow with some theoretical simulations and predictions on its convertibility toward graphene. In addition to that, we elaborate on extended chemical modifications (covalent and non-covalent), cross-linking, and doping recipes for this macromolecule shown in literature. After all, we intend to show you that GO became a relatively hot research topic, not only due to its relevance to graphene, but also for its high chemical activity and tunability, which enabled the prosperity of its research in various fields led by chemists, materials scientists, biologists, physicists, as well as engineers. It is a perfect paradigm for young researchers as an important subject thrived in interdisciplinary research. After all, when real-life problems come, potential solutions do not impose boundaries between disciplines. All relevant disciplines can offer their input, and contribute together to the final solutions, in which cases communications and collaborations between different researchers need to be encouraged and appreciated.

**Keywords** Graphene oxide • Reduction • Functionalization • Covalent • Non-covalent • Doping • Cross-linking • Toxicity • Hygroscopicity

## 3.1 Reduction

As mentioned in Chap. 1, the initial motivation of GO reduction is to produce graphene on a large scale. However, the reduction products of GO are relatively poor in crystallinity and hence carrier mobility, though they are incautiously claimed by many

W. Gao (✉)
The Department of Textile Engineering, Chemistry & Science, College of Textiles,
North Carolina State University, Raleigh, NC 27695, USA
e-mail: wgao5@ncsu.edu

© Springer International Publishing Switzerland 2015
W. Gao (ed.), *Graphene Oxide*, DOI 10.1007/978-3-319-15500-5_3

researchers as "graphene." In this chapter, we name them as reduced graphene oxides (rGOs), to distinguish them from CVD-grown or mechanically exfoliated graphene products. Some researcher would also like to call them chemically converted graphene (CCGs), which might be a more accurate term, since some chemical treatments do not involve any reducing reagent, and the reactions happened are basically disproportionation of GO itself. This section summarizes and compares different reduction recipes introduced for GO, in order to gain deeper understanding on reaction mechanisms with the help of modeling and theoretical analysis.

### 3.1.1 Comparison of Reduction Recipes

The most common reaction medium for GO is water, and there are different ways to disperse GO into water, including sonication and mechanical stirring. Sonication has been reported to create defects and decrease the sheet size of GO from several microns to few hundred nanometers, and also widen the size distributions, hence less favorable than mechanical stirring in many applications [1–3]. The dispersibility of GO in water is typically on the order of 1–4 mg/mL [4]. On the other hand, GO can also be dispersed in organic solvents such as dimethylformamide (DMF), *N*-methylpyrrolidone (NMP), and tetrahydrofuran (THF) [5, 6], and thus is also believed to be amphiphilic with the core more hydrophobic and edges more hydrophilic, acting just like a surfactant [7]. As a giant molecule with amphiphilicity, GO can be assembled into continuous single-layer films by the Langmuir–Blodgett method [8–10].

Independent of the dispersion medium, a stable colloidal dispersion of GO is readily reactive with a variety of chemicals, most of which have been reducing reagents. The first example was the hydrogen sulfide reduction introduced by *Hofmann* in 1934 [11]. Only one report on lithium aluminum hydride ($LiAlH_4$) reduction has come out [12], probably due to the strong reactivity of $LiAlH_4$ with the common dispersing media water. In their case, THF was used as a solvent and methanol was used to wash. They named the reduction product G–OH, indicating lots of –OH groups remaining on GO after $LiAlH_4$ treatment, and thus poor reduction effectiveness. On the other hand, although $NaBH_4$ is slowly reactive with water, the reaction is kinetically slow enough to allow GO reduction to happen; however, the strongly reducing $BH_4^-$ ions usually create a lot of defects on GO basal planes. So far, the most popular reductant for GO has been hydrazine [13]. The reduction mechanism was proposed at least for one of the functional groups on GO, as shown in Scheme 3.1 [13].

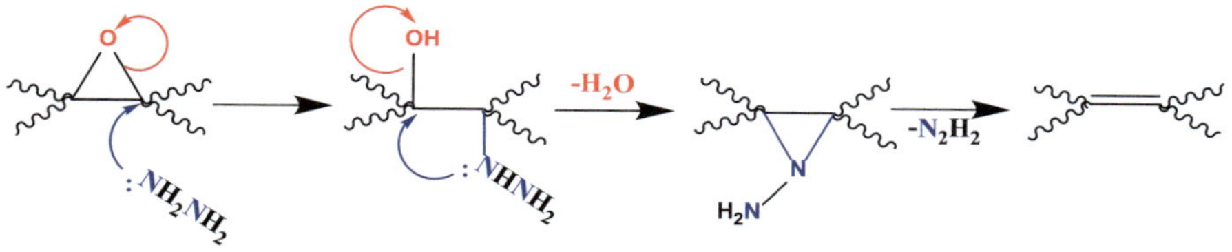

**Scheme 3.1** Reaction mechanism proposed for epoxide reduction by hydrazine

Besides these, GO was believed to be one of the most important precursors to graphene; thus in literature, lots of chemical/physical reduction protocols have been demonstrated, and to compare their effectiveness, characterizations of products with conductivity measurement, elemental analysis, $SS^{13}CNMR$, XPS, FTIR, Raman, XRD, TEM, near-edge X-ray absorption fine structure (NEXAFS) spectroscopy, etc. have been widely reported. Comparisons between different recipes have also been discussed [14–16]. Here we summarize most chemical and physical treatments and the product characterizations in Table 3.1, in order to provide a facile comparison of numerous methods demonstrated so far.

The restoration of π conjugation can be verified by changes between GO and rGO in UV–Vis spectra, XPS data, and electrical conductivities. The red shifts of UV adsorption peaks suggested the extension of π–π conjugation according to *Hückel*'s rule. The appearance of the π–π satellite peak in XPS was also a good indication. Electrical conductivity would be another good criterion to judge the degree of the restoration. As shown in Table 3.1, more than 30 chemicals have been reported as reducing reagent for GO, and so far the product with highest electrical conductivity is produced by hydroiodic acid with acetic acid treatment [40]. As mentioned before, we need to emphasize that all these chemical treatments worked as reduction protocols to GO, but their products were far from the firstly made graphene in their crystallinities [100], since the carrier mobility in these products is at least three orders of magnitude lower than that of graphene. In this case, we would like to conclude that GO as a chemically active compound is highly prone to reduction, but the defects and disordered structures in GO are very hard to be recovered. This can also be verified by Raman data in Table 3.1. All of those Raman spectra reported on rGOs have prominent D peaks, indicating highly defected structures in rGOs. In addition, most of them have higher D/G than that of GO, and researchers like to explain this higher D/G ratio by the increased $sp^2/sp^3$ carbon interfaces due to reduction, which seems a little bit farfetched. The argument against this explanation is that most of the highly reduced GO samples, where oxygen content is already pretty low (<0.5 wt%) [17], still have a higher D/G ratio than that in GO. Such a low oxygen content is just not enough to offer more $sp^2/sp^3$ carbon interfaces in the rGO sample. Theoretical modeling or simulations are needed for better understanding of this issue.

On the other hand, the reduction processes usually come with heteroatom incorporation into final products, further complicating structure and pushing rGOs further away from pristine graphenes. As shown in Table 3.1, both elemental analysis and XPS data show the existence of heteroatoms in final products, including oxygen, nitrogen, sulfur, boron, and hydrogen. Thus, people also tried to compare the purity of their products by comparing the ratio of C/O, C/(O + N), etc. These heteroatoms do influence the electrical conductivity as well, such as residual C–N groups can act as n-type dopants [74].

In order to elucidate the reduction mechanisms as well as we can, we have summarized how those functional groups on GO would react with various chemicals, as shown in Table 3.2. Most of these reactions are explained based on fundamental organic chemistry on the aspect of reactivity of individual functional

**Table 3.1** Comparison in reduction protocols of GO

| | Electrical conductivity (Siemens/cm) | C/O (elemental analysis) | XPS (eV) | Raman (cm$^{-1}$) | XRD (nm) | FTIR (cm$^{-1}$) | SS$^{13}$CNMR (ppm) | UV (nm) |
|---|---|---|---|---|---|---|---|---|
| GO | $5.3 \times 10^{-6}$ to $4 \times 10^{-3}$ [17, 18] | 2.7 [13] 2.44 [17] | 284.8 (C–C) 286.2 (C–O) 287.8 (C=O) 289.0 (C(O)–O) [13, 18] | 1,594 (G) 1,363 (D) [13] | 0.63–1.2 [13] | 1,060 (C–O) 1,220 (phenolic) 1,370 (OH bending) 1,620 (H$_2$O bending) 1,720 (C=O) [17] | 57 (C–O–C) 68 (C–OH) 130 ($sp^2$) 188 (C=O) [13] | 230 [19] |
| Hydrazine monohydrate [13, 20, 21] (NH$_2$NH$_2$·H$_2$O) | 2 | 10.3 | 284.5 with tails | 1,584 (G) 1,352 (D) | NA | NA | 117 ($sp^2$) | NA |
| Hydrazine monohydrate (98 %) [22] + leavening | Less than 100 Ω/sq. | NA | NA | 1,571 (G) 1,321 (D) D/G increased | NA | Decrease in absorption band intensities | NA | NA |
| Pure hydrazine [23] | 10$^8$-fold higher than GO | NA | 284.5 dominant 532 (O) 533 (O) after thermal annealing | 1,600 (G) 1,350 (D) 2,700 (2D) 2,950 (D+G) D/G increase | NA | NA | NA | NA |
| Vapor-phase hydrazine [24] | Four orders increase | NA | NA | Changes observed | NA | NA | NA | NA |
| Dimethylhydrazine [25] | $1 \times 10^{-3}$ with 1 vol.% in PS | NA | NA | NA | 0.426 0.245 | NA | NA | NA |
| p-Toluenesulfonyl hydrazide [26] | 1.64 | NA | 285.89 C–N | NA | NA | 1,052, 1,226, 1,727, 3,400 decreased | NA | 268 |

| | | | | | | | | |
|---|---|---|---|---|---|---|---|---|
| LiAlH$_4$ [12] | NA | 0.93 (48.25/51.75) | 531.1, 532, 532.7, 533.3, and 535 eV | NA | NA | NA | NA | NA |
| Trioctylphosphine [27] | 2.5 | NA | C/O=9.09<br>284.6 with tails | 1,608 (G)<br>1,312 (D) | 0.385 | 1,574 (C=C) | NA | 268 |
| Hydriodic acid with acetic acid [28] | 3.04×10$^2$<br>7.85×10$^3$ (vapor phase) | 15.27 | C/O=6.67<br>284.6 with tails | 1,581 (G)<br>1,350 (D)<br>D/G=1.10 | 0.362 | Absent of obvious peaks | 110.1 ($sp^2$) | Flat absorbance up to 900 |
| Hydrogen iodide (HI) [29] | NA | NA | NA | NA | NA | 1,710 (C=O) disappeared | NA | NA |
| Iodide (HCl+KI) [30] | 12.51 | NA | 284.5 with tails | 1,575 and 1,350<br>D/G=1.32 | 25.93° | Dramatic decrease in the peak intensities | NA | 264 |
| Melatonin [31] | NA | NA | 284.5 with tails | 1,583 (G)<br>2D/G=0.23<br>D/G=1.07 | NA | NA | NA | 269 |
| Deionized water [32] | Five orders of magnitude decrease | 6 | 284.4 with tails | NA | None | Reduction in C=C, C=O, and C–O–C | NA | NA |
| Hydrothermal steam etching [33] | Increase observed | NA | 287 peak decrease | NA | NA | NA | NA | NA |
| Aqueous phytoextracts [34] | 40.46 | 7.11 | NA | 1,584 (G)<br>1,327 (D)<br>2,709 (2D) | 0.36 | Disappearance of peaks | NA | 268 |
| SO$_2$ [35] | NA | NA | C/O=4.49 | 1,583 (G)<br>1,349 (D)<br>D/G=1.01 | 0.395 | NA | NA | 272.5 |

(continued)

**Table 3.1** (continued)

| | Electrical conductivity (Siemens/cm) | C/O (elemental analysis) | XPS (eV) | Raman (cm⁻¹) | XRD (nm) | FTIR (cm⁻¹) | SS$^{13}$CNMR (ppm) | UV (nm) |
|---|---|---|---|---|---|---|---|---|
| Hydrazine with NH3 | 4.88 | 15.06 | 67.9 % (sp2) | NA | 0.368 | NA | NA | NA |
| (NH)/HI in acetic acid (HI); (HI/NH) [36] | 24.5 | 16.58 | 75.6 % (sp2) | | 0.368 | NA | NA | NA |
| Li metal in liquid ammonia | NA | NA | 284.8 with tails | 1,854<br>1,352 | 26° (2θ) | NA | NA | NA |
| Lithium naphthalenide [37] | 4.24 | 8.86 | NA | 1,605 (G)<br>1,357 (D) | 24⁰ (2θ) | Peak intensities decreased | NA | NA |
| Poly(diallyldimethylammonium chloride) [38] | NA | 15.5 % for O, atomic percent | 285.9 for C–N | NA | 20° (2θ) | 1,385 (N–O)<br>831 (C–N) | NA | NA |
| Hydraulic acids (HI) [39, 40] | 298 | 12 | 284.6 with tails | 2D (2,467) increases<br>D/G increases<br>165 for polyiodides | 0.357 | NA | NA | NA |
| Hydroxylamine [41] | 11.22 | 9.7 | 284.8 | D/G increases | 0.372 | All peaks decrease | NA | NA |
| Benzylamine [42] | NA | 4.77<br>10.21 (2 h reduction) | C–N appear<br>C–O and C=O decrease | 1,594 (G)<br>1,350 (D) | NA | NA | NA | 260 |
| Dopamine [43] | NA | NA | 285.5 for C–N | NA | 0.381 | NA | NA | 268 |
| Nonaromatic amino acids [44] | 7 | NA | NA | D/G=1.11 | No peak | Peaks almost disappeared | NA | NA |
| Ethanethiol–aluminum chloride complexes [45] | NA | 4.71 | Prevalent C=O and O–C=O peaks | D/G=1.07 | NA | O–H stretching disappeared, C=O lowered | NA | NA |

| Method | | | | | | | | |
|---|---|---|---|---|---|---|---|---|
| Microwave [46–48] | 2 | 4.5 | 284.5 with tails, $\pi$–$\pi^*$ bump | 1,591 (G)<br>1,348 (D)<br>D/G=0.96 | 0.355/no peak | 1,562/1,577 appear<br>1,724 decrease<br>1,622 absent | NA | 260 |
| Methane plasma [49] | 9.0 k$\Omega$/sq.<br>1,590 | NA | C–O and C=O decrease | D/G=0.53 | NA | NA | NA | NA |
| Sodium hydrosulfite ($Na_2S_2O_4$) [50] | 13.77 | NA | 79 % (C–C, C–H) | D/G=1<br>1,570.7 (G)<br>1,347.3 (D) | 0.377 | NA | NA | NA |
| Polyphenol [51] | $4.33 \times 10^1$ | NA | 284.6 with tails | D/G=1.18 | 0.43 | NA | NA | 278 |
| Supercritical alcohols [52] | NA | 10.4–14.4 | 1.15 % carbonyl | D/G=1.30 | 0.358–0.381 | Peaks disappeared | NA | NA |
| NaBH$_4$ [19] | $1.5 \times 10^{-6}$–$4.5 \times 10^1$ | 2.6–8.6 | 284 with tails<br>$\pi$–$\pi^*$ bump | D/G increase upon $C_{NaBH4}$ increase | 0.380 to 0.373 | NA | ~120 ($sp^2$) | 260 |
| Variable-valence metal-assisted NaBH$_4$ [53] | $3 \times 10^2$ | 1.81–4.99 | 284.6 with lower tails | D/G:<br>1.39–1.48<br>1,595 (G)<br>1,360 (D) | 0.356–0.366 | All disappear except for 1,220 (C–OH) | NA | NA |
| Thermal reduction [54] | 0.009–2.75 | NA | NA | D/G increase | NA | NA | NA | NA |
| Single large-sheet thermal annealing [55] | 760 | NA | Decrease of oxygenated carbon peaks<br>89 % C–C | D/G decrease<br>1,586 (G) | NA | NA | NA | NA |
| Solvent thermal reduction [56–58] | 2.3/52.30 | NA | C/O=4.70<br>C/O=6.8–8.3 | 1,586 (G)<br>1,347 (D) | 0.41/0.36 | 1,573 exists/Most peaks disappeared | NA | NA |
| Hydrothermal dehydration [59] | NA | NA | $sp^2/sp^3$=5.6 | 1,593 (G)<br>1,352 (D)<br>D/G=0.90 | NA | NA | Broad 94–160 | 254 |

(continued)

**Table 3.1** (continued)

| | Electrical conductivity (Siemens/cm) | C/O (elemental analysis) | XPS (eV) | Raman $(cm^{-1})$ | XRD (nm) | FTIR $(cm^{-1})$ | SS$^{13}$CNMR (ppm) | UV (nm) |
|---|---|---|---|---|---|---|---|---|
| Sulfur-containing compounds (NaHSO$_3$, Na$_2$SO$_3$, Na$_2$S$_2$O$_3$, Na$_2$S·9H$_2$O, SOCl$_2$, and SO$_2$) [60] | $6.5 \times 10^1$ (NaHSO$_3$) | 6.48–7.89 (NaHSO$_3$)<br>2.32 (Na$_2$SO$_3$)<br>3.88 (Na$_2$S$_2$O$_3$)<br>5.61 (Na$_2$S)<br>6.49 (SO$_2$)<br>6.75–8.48 (SOCl$_2$) | 284.7 (C=C)<br>285.5 (C–C) | 1,352 (D)<br>D/G: 0.95–1.22 | NA | NA | 1,577 (C=C, aromatic) | NA |
| Diethylaminosulfur trifluoride [61] | 398 Ω/sq. | 9.24 | 289.7 (CF$_2$) | D/G = 1.63 | 0.41 | 1,200 (CF$_2$) | NA | NA |
| Oligothiophene [62] | NA | 6.12 | 286.2 (C–S) | D/G: 1.02, 1.11, 1.18<br>1,595 (G) | NA | C=O and C–O decreased | NA | NA |
| Formamidinesulfinic acid [63] | 2.9 | 6 | 284.6 with tails | 1,596 (G)<br>1,344 (D)<br>2,688 (2D) | 22.39° | Peak intensity decreased | NA | 264 |
| Vitamin C [64–68] | 8<br>$7.7 \times 10^1$<br>0.141<br>15 | 12.5 | 284.5 with tails<br>284.6 (C–C) fwhm 0.8–1.1<br>N content observed<br>284.5 with tails | D/G increase<br>D/G = 1.752 | 0.37 | 1,726, 3,395, 1,410, 1,226, 1,025 decrease dramatically<br>1,300–1,350 | NA | 264/268 |

|  |  |  |  |  |  |  |  |  |
| --- | --- | --- | --- | --- | --- | --- | --- | --- |
| Wild carrot root [69] | NA | 11.9 | NA | 1,582 (G)<br>2,700 (2D) appears | 23.96° (2$\theta$) | C=O and O–H stretches decrease | NA | NA |
| *Hibiscus sabdariffa* L. [70] | 36.50 $\Omega$/sq. | 3.1 | Small peaks between 286.1 and 289.4 | 1,586 (G)<br>1,326 (D)<br>D/G = 1.24 | 0.36 | Peak intensities reduced | NA | 268.5 |
| Bacteriorhodopsin | $7.1 \times 10^4$ $\Omega$/sq. | >5 | 285.9 (C–N) | D/G = 0.08<br>Decreased | NA | NA | NA | NA |
| KOH/NaOH [71] | NA | NA | 291.5 ($\pi$–$\pi$*) | NA | NA | NA | 90–150 | NA |
| Reducing sugar [72] | NA | NA | Oxygen-binding peaks decrease | 1,584 (G)<br>1,354 (D)<br>D/G increase | No peaks | Peaks from oxide groups decrease | NA | 261 |
| Bovine serum albumin [73] | NA | NA | 284.6 with tails | NA | NA | NA | NA | 268 |
| $H_3PO_4/H_2SO_4$ [74] | 69 | NA | C/O = 8.5–1<br>1.7<br>291.5 ($\pi$–$\pi$*) | D/G:0.85 | NA | C–H (2,950)<br>C=C (1,600) | NA | 270 |
| $NaBH_4 + H_2SO_4 +$ thermal annealing [17] | $2.02 \times 10^2$ | >246 | 284.5 with tails | 1,582 (G)<br>1,346 (D) | 0.337 | No signal | 119 (CCG2)<br>105 (CCG3) |  |
| Electrochemical reduction [14, 59, 75–81] | 35<br>85 [76] | 23.9 [76] | NA | 1,595 (G)<br>1,360 (D)<br>2D observed | 0.335 | C–O remains | NA | NA |
| Electrochemical reduction (tunable) [82] | NA | 3–10 | Oxidized carbon signal decreases | NA | NA | NA | NA | NA |

(continued)

**Table 3.1** (continued)

| | Electrical conductivity (Siemens/cm) | C/O (elemental analysis) | XPS (eV) | Raman (cm$^{-1}$) | XRD (nm) | FTIR (cm$^{-1}$) | SS$^{13}$CNMR (ppm) | UV (nm) |
|---|---|---|---|---|---|---|---|---|
| Galvanic displacement [83] | NA | 11.78 (Cu/Ag) 7.46 (Cu/Au) | NA | D/G: 0.90 (Cu/Ag) 0.83 (Cu/Au) 1,585 (G) | NA | Peaks almost vanish | NA | NA |
| Al powder [84] | $2.1 \times 10^1$ | 18.6 | 284.6 with tails | D/G=1.81 | 0.375 | NA | NA | NA |
| Zn power with ultrasonication [85] | 150 | 33.5 | Close to graphite, 284.5 eV with tails | D/G=~1.8 G (1,580) | NA | Most peaks weakened or disappeared | NA | Red shift to 272 |
| Hydroquinone [86] | NA | NA | NA | 1,595 (G) 1,350 (D) | NA | NA | NA | NA |
| Hydrogen (H$_2$) [87, 88] | $1 \times 10^3$ | NA | 284.3 with tails C/O:10.8–14.9 | NA | NA | NA | NA | NA |
| UV irradiation [89–91] | NA/One order decrease | NA | C/O=10 | D/G decrease | NA | NA | NA | Red shift of adsorption peak |
| VUV synchrotron radiation [92] | NA | C–O/C–C: 0.60 and 0.65 | NA | NA | NA | NA | NA | NA |
| Flash light [93] | 10 | 4.23 | NA | NA | 22.5° | Decrease of major peaks | NA | NA |
| Hydrogen-assisted pulsed KrF-laser irradiation [94] | 500 Ω/sq. | 90 % decrease in O% | 284.8 with tails | 2D appear | No peak | C=C and C=O unchanged | NA | NA |
| Laser irradiation [95] | NA | NA | NA | 1,620 (D') 1,586 (G) 1,331 (D) | NA | NA | NA | NA |

| Pulsed laser [96] | 53.8 kΩ/sq. | NA | 8 % C–O<br>7.6 % C=O | 1,584 (G)<br>1,327 (D)<br>D/G = 1.08 | No peak | More peaks disappeared | NA | Red shift |
|---|---|---|---|---|---|---|---|---|
| Gamma-ray irradiation in gaseous phase [97] | 23.34 | NA | NA | D/G = 0.89 | 24° (2$\theta$) | NA | NA | NA |
| Hot-pressing [98] | 1,000 cm$^2$ V$^{-1}$ s$^{-1}$ (electron mobility) | NA | 284.7 with tails | D eliminated 2D recovered at 40 MPa | NA | NA | NA | NA |
| Ball-milling with H$_2$ [99] | 34 | 6.17 and 6.80 | Peak intensities decreased | 1,598 (G)<br>1,357 (D) | 0.343 | 1,590 (C=C) | 115 ppm | NA |

**Table 3.2** Reactivity of functional groups on go with different chemical reagents

| | $NH_2NH_2$ | $NaBH_4$ | $LiAlH_4$ | KOH/NaOH [48] | $H_2SO_4$ (conc.) [101] | $H_3PO_4$ [68] | HI and AcOH [40] | Diazonium salts [102] |
|---|---|---|---|---|---|---|---|---|
| Epoxy (C–O–C) [29] | $C=C/C_2N_2H_2$ | CH–COH (low) | CH–COH | COH–COH | NA | $COH–CH_2PO_4$ | CI–COH | NA |
| Hydroxyl (C–OH) | $CNHNH_2$ | NA | NA | NA | C=C | NA | C–I | NA |
| Ketone (C=O) | $C=NNH_2$ | C–OH | CHOH | COH–COH | NA | $COH–CH_2PO_4$ | CICOH | NA |
| Carboxyl (–COOH) | inert [48] | –NA | –CHOH | –COONa | NA | NA | NA | NA |
| $sp^2$ Carbon | –NA | –NA | NA | –NA | –NA | NA | NA | $C(sp^3)$–Ph– |
| Ester (–O–C=O) | $O–C=NNH_2$ | NA | –O–CHOH | –COOH | –NA | –COOH | NA | NA |
| Lactol (–O–C–OH) | $O–C–NHNH_2$ | NA | –O–CHOH | NA | –NA | NA | NA | NA |

groups. As we can see from the table, even the most popular reagent "hydrazine" would still leave some functional groups intact. Therefore, though the $sp^2$ carbon lattice structure is thermodynamically favorable over GO, the complete restoration of $\pi$ conjugation is really hard, let alone the refilling of carbon vacancies that have been created in the harsh oxidation processes.

### 3.1.2   Theoretical Simulations and Predictions

Early in 2008, Boukhalov et al. modeled the chemical structure of GO, and the process of its deoxygenation [103]. They claimed that $C_8(OH)_4O$ is the most energetically favored composition for GO, and the coexistence of –OH and epoxy groups is also highly possible. It is relatively easy to convert $C_8(OH)_4O$ to $C_{32}(OH)_2$, since the energy difference between the two is less than 1 eV. But it is really hard to push the reduction further. As shown in Fig. 3.1, the authors believe that most of the chemical reductions

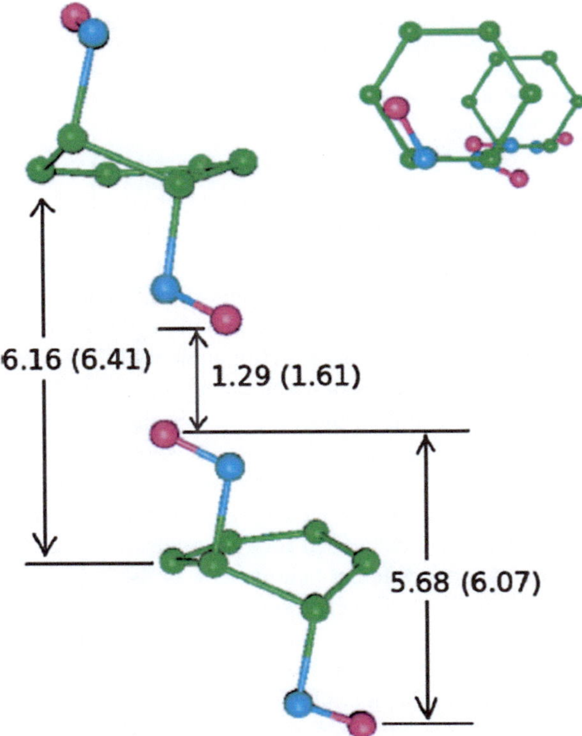



**Fig. 3.1** Optimized geometric structures of strongly reduced GO. *Numbers* are distances, in Å, for the periodic structure (and for bilayer in *parentheses*). *Right upper corner*: a top view. Carbon, oxygen, and hydrogen atoms are shown in *green*, *blue*, and *violet*, respectively.. (Adapted with permission from Ref [103]. Copyright (2008) American Chemical Society)

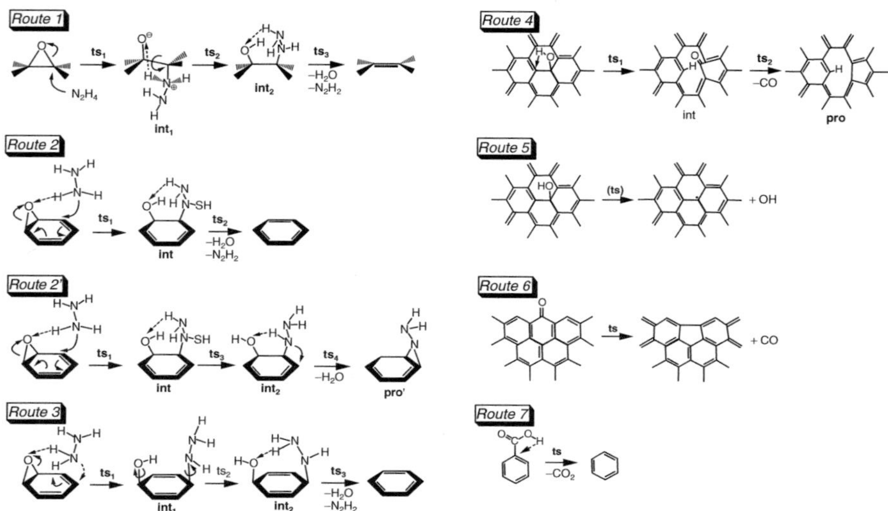



**Fig. 3.2** GO reduction mechanisms. *Routes 1–3* and *2'* represent the mechanisms for the hydrazine de-epoxidation of GO. *Routes 4* and *5* refer to the mechanism for the thermal de-hydroxylation of GO. *Routes 6* and *7*, respectively, show the mechanisms for thermal de-carbonylation and thermal decarboxylation of GO. (Adapted with permission from Ref. [104]. Copyright (2010) American Chemical Society)

will lead to the more stable product with an interlayer coupling energy of 17 meV, much lower than the 35 meV for graphene. Because of this weak coupling, the electronic structure of GO or partially reduced GO is almost the same for single layer and multiple layers, in contrast to the cases of mechanically exfoliated graphenes [103].

In 2009, Gao et al. investigated the deoxygenation process of GO with hydrazine or heat treatment based on density functional theory [104]. They found in both treatments that oxygenated moieties on basal planes are easier, both kinetically and thermodynamically, to remove than those on the peripheries. Hydrazine reduction tends to stay at the form of hydrazino alcohols. As shown in Fig. 3.2, four routes (routes 1, 2, 2', and 3) were proposed for hydrazine reduction, and another four routes (4–7) were proposed for thermal annealing. Energy profiles (relative enthalpies and Gibbs free energies) were calculated for each route. The prediction of residual functional groups in GO after different treatments was also made available.

Later in the year, gradient-corrected spin-polarized density-functional theory was used by Kim et al. to further clarify the reaction mechanisms for epoxide reduction with hydrazine [105]. Eley–Rideal mechanism was favored over Langmuir–Hinshelwood mechanism for the epoxide reduction. In detail, epoxide is reduced to form $H_2O$ by two successive hydrogenation reactions while two hydrogen atoms come from hydrazine molecules or its derivatives. However, as noted by the authors, the calculation they did was for vapor phase, whereas most reduction of GO with hydrazine happens in aqueous solutions. Further clarifications for solution-phase reactions are needed.

Whereas theoretical modeling and simulations or even predictions greatly helped GO chemists understand the structure and reduction pathways for GO, the matching between theoretical predictions and real experimental experiences is still hard to obtain. The authors of this book are eager to see more efforts in both theory and experiments, so that thorough understanding and accurate modeling can be demonstrated along with various experiments. It would be very interesting to see experimentally that intermediates in different reaction pathways predicted are separated and analyzed to ultimately prove the theoretical profiles proposed.

## 3.2  Functionalization

The addition of other functional groups directly onto GO, forming either covalent or non-covalent attachments, falls into another big category of GO chemistry. Due to the ambiguity of GO structure, we emphasize here that functionalization chemistry of GO is complex and not well understood. According to the widely accepted *Lerf–Klinowski* model, reactive functional groups on GO are epoxy, hydroxyl, organosulfate, carbonyl, carboxyl, and ester moieties. It would be nice if one can selectively react with one of these groups and keep others intact; however, so far such kind of orthogonal reactions have not been demonstrated. Most of the functionalizations occurred on more than one type of oxygenated groups and resulted in very complicated products with separation and purification almost impossible. Regardless of vague chemical characterizations, most of these products showed interesting applications in various fields.

### 3.2.1  Covalent

Among different functional moieties, carboxylic acid might be the most active one, since it is mostly located on the peripheries of GO sheets. The activation the –COOH is usually led by treatment with $SOCl_2$, followed by various nucleophilic attacks with different nucleophiles (Fig. 3.3) [107–110].

1-Ethyl-3-(3-dimethylaminopropyl)-carbodiimide (EDC) [111], *N,N'*-dicyclohexylcarbodiimide (DCC) [71], or 2-(7-aza-1H-benzotriazole-1-yl)-1,1,3,3,-tetramethyluronium hexafluorophosphate (HATU) [8] can also activate –COOH, and attacks by nucleophiles such as amines or hydroxyl groups are usually followed in order to form covalent attachments. The resulted amides have shown possible applications in optoelectronics [108, 110, 112, 113], catalysis [114], biodevices [8, 115], drug-delivery vehicles [111, 116], supercapacitors [117], and polymer composites [118–120]. A more complicated case involves the attachment of diamine, further covalently binding with bromide-terminated initiators, and subsequent polymerization on GO surface [118]. The resulted GO/polymer composites usually offer better dispersibility in many solvents. Besides that, isocyanate derivatives have also been shown to react with carboxyl and hydroxyl groups leading to

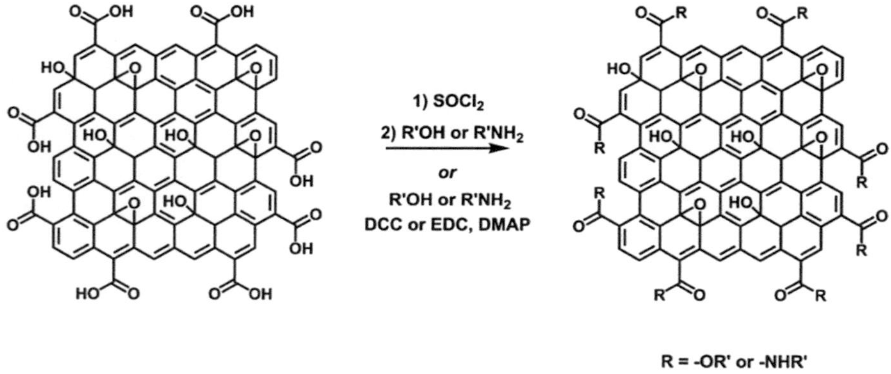

**Fig. 3.3** Activation of GO's peripheral carboxylic acid groups with either SOCl$_2$ or a carbodiimide, and subsequent condensation with an alcohol or an amine. (Reproduced from Ref. [106] with permission of The Royal Society of Chemistry)

amide and carbamate esters [121]. The products here can be well dispersed in polar aprotic organic solvents. Chitosan chains also reacted with the carboxyl groups on GO-forming amino bonds with only microwave assistance, offering possible biomedical applications [93].

Epoxy group is another major functionality on GO. The epoxy rings can be easily opened under acidic conditions or by nucleophilic attack. For example, octadecylamine was used to react with GO and offered a colloidal dispersion of functionalized GO (FGO) in organic solvents [122]. Hexylamine was also used to attack the epoxy rings on GO to form alkylated GO which can further be reduced into alkylated conductive graphene paper [74]. Ethylenediamino-β-cyclodextrin was also introduced to attach cyclodextrin onto GO surface via amine-epoxy reaction (Fig. 3.4) [123]. An ionic liquid (1-(3-aminopropyl)-3-methylmidazolium bromide, RNH$_2$) was attached to GO via the end amine group in a nucleophilic attack of the epoxy groups [124]. Another example involved 3-aminopropyltriethoxysilane (APTS) attachment onto epoxy via a S$_N$2 reaction and of course ring opening of epoxy groups, while reinforcement in the mechanical properties of the resulted silica composite was demonstrated [125]. Interestingly, inspired by the synthetic polymer chemistry or biochemical systems, cross-linking of GO with poly(allylamine) or sodium borate, via epoxy and hydroxyl groups, has also been investigated [126, 127], and mechanical enhancement of the resulted GO film was observed in both cases. Unfortunately these cross-linking strategies cannot stabilize GO in solvents especially in water, unlike the case for polymers, where cross-linked GO paper still breaks down when in contact with water, and hence other chemical methodologies are needed to tackle this problem.

Hydroxyl groups on GO can act as nucleophiles to attack ketones. For instance, 2-bromo-2-methylpropanoyl bromide was used to react with hydroxyl groups on GO to form an initiator for atom transfer radical polymerization (ATRP), offering GO-polymethyl methacrylate (GO-PMMA) as the final product (see Fig. 3.5) [128].

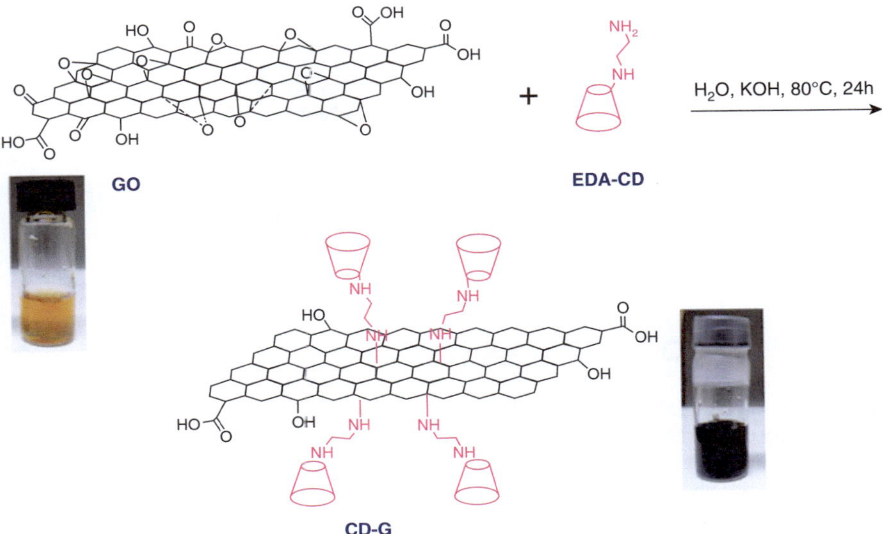



**Fig. 3.4** Synthetic route of CD-G and the corresponding photos of aqueous solutions of GO and CD-G. (Reprinted with permission from Ref. [123]. Copyright (2011) American Chemical Society)

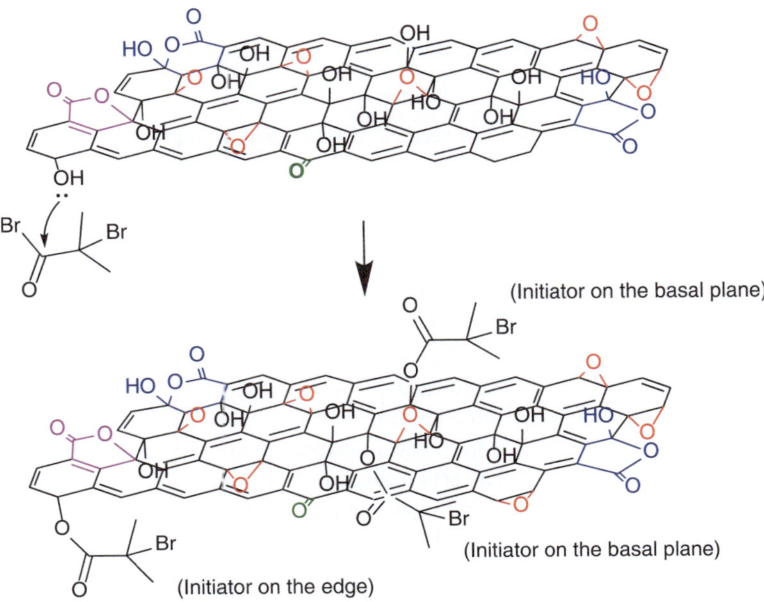



**Fig. 3.5** Nucleophilic attack of hydroxyl groups on GO to 2-bromo-2-methylpropanoyl bromide, offering an initiator for ATRP

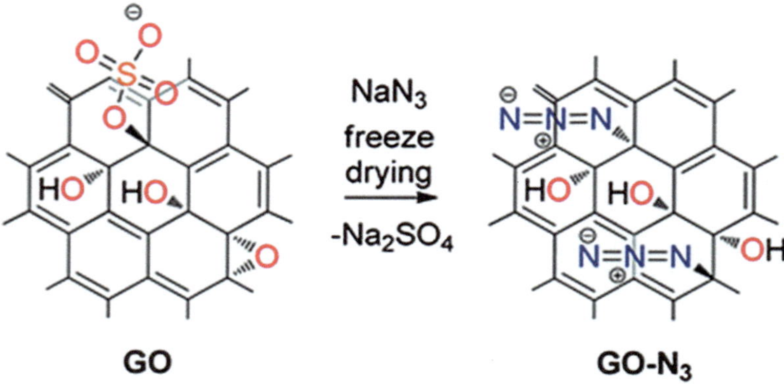



**Fig. 3.6** Reaction of GO with $NaN_3$ in solid state. (Adapted from Ref. [130], published by The Royal Society of Chemistry)

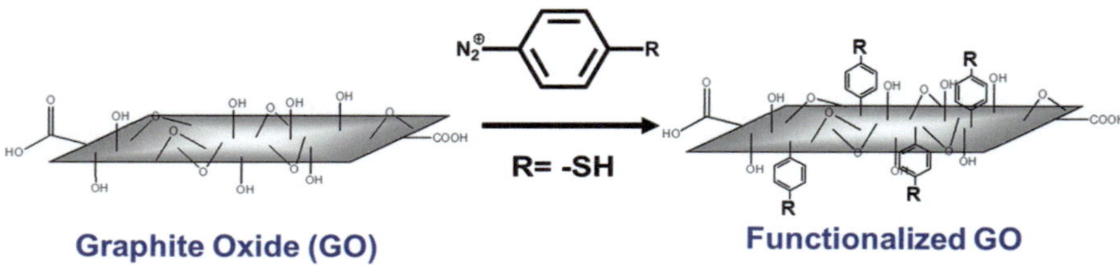



**Fig. 3.7** Diazonium salt functionalization of GO. (Reprinted with permission from Ref. [133]. Copyright (2011) American Chemical Society)

Another interesting example is to react hydroxyl groups with nitriles by McGrail et al. [129]. The advantages of their recipe include the following: (1) the reaction can be done easily and rapidly in aqueous solution with a variety of small molecules and polymer nitriles; and (2) the products can be separated with centrifugation or filtration, and maintain tunable solubility and functionalities.

As we mentioned in Chap. 1, a small amount of organosulfate groups are present in as-synthesized GO. These moieties have been replaced with azide anions in solid state in Eigler's report (Fig. 3.5) [130]. The biggest advantage of this reaction recipe is that the thermally unstable functional groups in GO, such as hydroxyls and epoxides, are preserved in the mild reaction conditions.

It is also worthwhile to mention the covalent functionalization of rGOs with diazonium salts [4, 131, 132]. The aryl diazonium salt was believed to react with the $sp^2$ carbon domain in rGOs [56] and yield highly functionalized rGOs with superior dispersibilities. Direct reacting of diazonium salts with GO was also reported by Gao et al. [133] (Fig. 3.7), rendering thiophenol functionalized GO with high mercuric ion adsorption capability. Another example of $sp^2$ carbon reaction in GO was shown by Ballesteros-Garrido et al., where phenothiazinyl units were attached to graphitic carbon via nitrene insertion [134]. All these reactions mentioned above lead to strong covalent bonding between GO and the other chemical, and most of them happen with more than one type of functionalities on GO, proving GO to be a very active compound. However, also due to the complex environment on GO surfaces, especially the existence of long pairs of electrons on most oxygen atoms, the actual reaction pathways are highly complex, and the covalent attachments of functional molecules to $sp^2$ carbons are yet to be verified. So far, even with the powerful solid-state NMR techniques, the formation of this type of covalent bonds is still hard to be evidenced. Therefore, although these literatures claimed such type of chemistry on GO, further investigations are desired to fully understand the reaction details on these giant molecules.

### 3.2.2 Non-covalent

Additionally, non-covalent functionalization of GO has also been demonstrated. Non-covalent interactions such as hydrogen bonding, $\pi$–$\pi$ stacking, cation-$\pi$, or van der Waals interactions mainly happen on oxidized or $sp^2$ carbon domains of GO. The strength of a $\pi$–$\pi$ interaction depends on the number of aromatic rings in the $\pi$-electron system of adsorption molecules and on the contact curvature between the aromatic region of the molecule and the surface of graphitic materials. The activation energy of this process is reported to be higher than that of physical adsorption processes (5–40 kJ mol$^{-1}$) [135]. For instance, 0.7 wt% of GO in poly(vinyl alcohol) (PVA) resulted in a 76 % increase in tensile strength and a 62 % increase in Young's modulus, both of which are due to the hydrogen bonding formed between PVA polymer chains and functional groups on GO [136]. Another example involves silk fibroin with pristine GO flakes introduced by Tsukruk and coworkers [137]. A spin-assisted layer-by-layer technique was used to fabricate the GO/fibroin composite membranes with varying ratio of GO (e.g., 3.0, 6.0, 9.0, 11.5, and 23.5 vol.%), and a Young's modulus as high as 145 GPa was obtained with 23.5 vol.% of GO, as compared to 10 GPa for original fibroin. Fibroin is a protein found in silk, which usually contains both hydrophilic and hydrophobic regions. The authors believe that the synergy of multiple non-covalent interactions between fibroin and GO is the major reason for the enhanced mechanical properties (Fig. 3.8) [137].

A GO-based biosensor has been demonstrated utilizing the GO-protein/DNA $\pi$–$\pi$ interactions [111]. Doxorubicin hydrochloride was also reported to form hybrid with GO via non-covalent interactions [138]. Sulfonated poly(ether-ether-ketone) was also used to non-covalently functionalize GO and the resulted product showed increased Young's modulus by 160 % and decreased oxygen permeability by 91 % as compared to pure poly(vinylidene fluoride) [139]. Sulfonated polyaniline was used to improve the water dispersibility of rGOs [140], where $\pi$–$\pi$ stacking and cation-$\pi$ interactions are probably involved, and high conductivity, good electrocatalytic activity, and stability were also obtained. $\pi$–$\pi$ stacking of pyrene molecules with a functional segmented polymer chain onto rGO surfaces resulted in a remarkable improvement in the thermal conductivity of graphene nanosheet-filled epoxy composites [141].

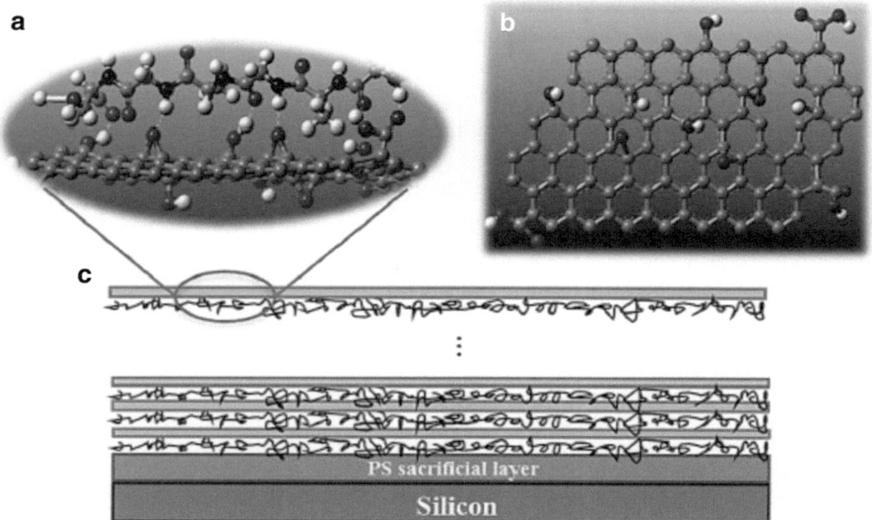


**Fig. 3.8** Molecular models of silk fibroin hydrophilic segments with polar interactions (**a,** *upper part*) and GO (**a**, *bottom part*: side view; **b**: top view). Elements in the *ball-and-stick model* are *grayscale* coded: H: *white*, C: *light gray*, O: *dark gray*, N: *black*. (**c**) The layered structure of nanocomposite silk fibroin-graphene oxide membrane. (Adapted from Ref. [137]. Copyright (2014) John Wiley & Sons Inc.)

## 3.3   Cross-linking

As expected, freestanding GO films can offer numerous benefits such as easy handling; certain mechanical strength; thermal, electrical, and ionic conductivities via modifications; water permeability; catalytic activity; biocompatibility; and light sensitivity. However, GO films only maintain less than 10 % of the stiffness and less than 1 % of the ultimate tensile strength of that in an individual graphene oxide sheet [142], mainly due to the predominate hydrogen-bonding interactions in between stacked GO sheets. Furthermore, GO films are susceptible to water. One drop of aqueous solution can easily lead to their deformation or even disintegration. Therefore, to fully utilize all possible functions of GO films and its composites, effective cross-linking of GO is highly demanded. The first cross-linking recipe introduced was with poly(allylamine) (PAA) [126] in 2009, in which Park et al. [126] and later on Satti et al. [143] produced composite GO/PAA papers with enhanced tensile strength up to 146 MPa.

Starting in 2011, several bio-inspired cross-linking recipes emerged in literature [127, 144–146]. The first inspiration offered by An et al. was from borate ions in higher order plants [127]. ~0.001–0.01 dry wt% [147] of borate ions present in those plant tissues enhance their mechanical properties, via covalent bonding with oxygen-containing functional groups in a variety of pH and counterion environments.

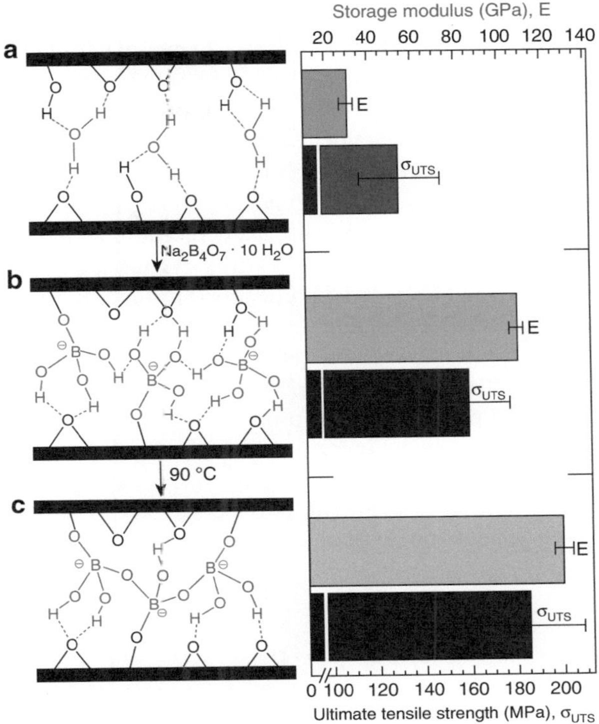


**Fig. 3.9** *Left*: Schematic illustration of the formation of the borate-cross-linked network across two adjacent graphene oxide nanosheets in a thin film. (**a**) Water molecules form hydrogen bonds with epoxide and hydroxyl groups to bridge nanosheets in an unmodified graphene oxide thin film. (**b**) In addition to forming hydrogen bonds, borate anions react with surface-bound hydroxyl groups to give borate orthoester bonds, improving the mechanical properties beyond those of the unmodified film. (**c**) Thermal annealing drives the formation of more covalent bonds within the intersheet gallery, affording an ultra-stiff thin film. *Right*: Mechanical properties of respective films, demonstrating that ultra-stiff films are only obtained after annealing borate-modified films under a flow of dry $N_2$. (Adapted from Ref. [127]. Copyright (2011) John Wiley & Sons Inc.)

In the case of GO, the authors believed that the borate ions covalently bound with the hydroxyl groups on GO, which resulted in 255 % and 20 % increase in stiffness and strength, respectively (Fig. 3.9) [127].

Another interesting inspiration from nacre came out in 2012 by Li et al. [144], where they mixed GO dispersion with poly(vinyl alcohol) (PVA) followed by reduction with hydroiodic acid (HI). They obtained a nacre-like lamella structure with improved mechanical properties and high electrical conductivity. Cheng et al. got similar inspiration from nacre, while in their case, 10,12-pentacosadiyn-1-ol was used as the cross-linking agent [145]. The advantage of this recipe is the decreased polymer content and improved tensile strength and toughness. Tian et al. demonstrated the use of dopamine, inspired by the adhesive proteins of mussels [146].

With the fine-tuning of pH in their systems, they obtained a cross-linked GO/polyetherimide (PEI) composite film with 550 % increase in Young's modulus over pristine GO films [146].

Besides bio-inspirations, several pure chemical protocols also emerged around the same time [148–150]. Glutaraldehyde/resorcinol [148], $Fe_3O_4$ nanospheres [149], and poly(oxypropylene) diamines [150] have been exploited in these studies, and the products show interesting applications in $CO_2$ storage, lithium ion batteries, and biomedical systems. In addition to that, many recent reports have been focused on GO-based aerogels and hydrogels [151–157]. As for aerogels, freeze-drying [151], supercritical drying [152], or lyophilization [153] are typically used to produce aerogel monoliths with extremely low density (0.16–60 mg/cm³) (Fig. 3.10). Several cross-linking reagents, such as CNTs [151], $La(OH)_3$ [152], polyethylenimine (PEI) [152], and ethylenediamine (EDA) [153], were used in these systems to improve the mechanical properties of the products.

Unlike aerogels, hydrogels are cross-linked networks that absorb large quantities of water without dissolving and are traditionally formed by physical or chemical cross-linking of natural or synthetic polymers [158]. Hydrogels made from GO or rGOs show interesting mechanical and electrical properties [154–156].

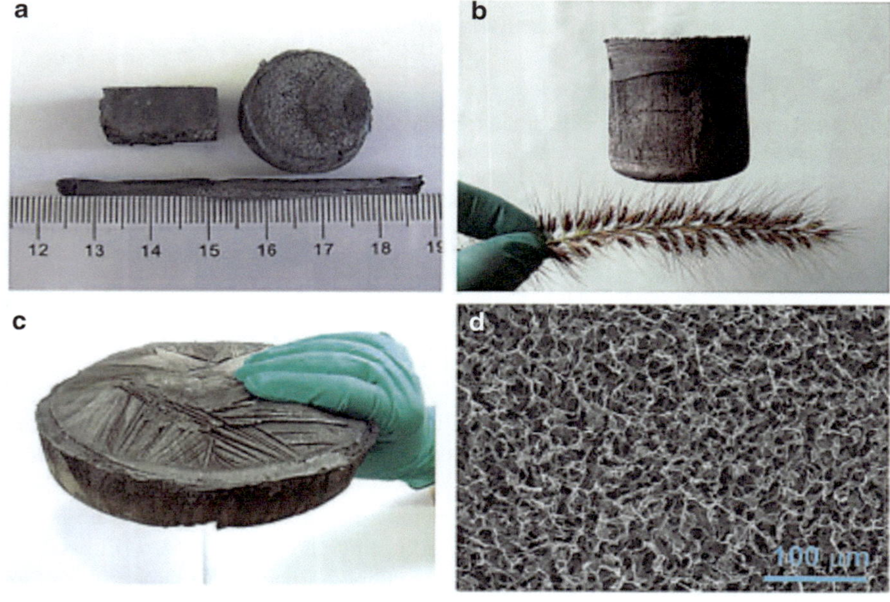

**Fig. 3.10** Macroscopic and microscopic structures of ultra-flyweight aerogels (UFAs). (**a**) Digital photograph of UFAs with diverse shapes. (**b**) A 100 cm³ UFA cylinder standing on a flower like dog's tail (*Setaria viridis* (L.) Beauv). (**c**) A ~1,000 cm³ UFA cylinder (21 cm in diameter and 3 cm in thickness). (**d**) Microscopic SEM image of a UFA showing CNT-coated graphene cell walls. (Adapted from Ref. [151]. Copyright (2013) John Wiley & Sons Inc.)

Chemical reduction or cross-linking with polyacrylamide [155] or $N,N'$-methylenebis(acrylamide) [156] was involved in order to make these hydrogels more conductive or elastic. Furthermore, external magnetic field was introduced to force the anisotropic alignment of GO flakes in the hydrogel [156], and the following cross-linking was able to retain the orientation of GO flakes even in the absence of magnetic field. This work has shed light on the desired manipulation of the group of 2D materials in closely related field.

In summary, a variety of cross-linking recipes have been introduced for GO and rGOs, resulting in mechanically enhanced membranes or composites with many promising applications. However, as mentioned in the beginning of this section, one of the major concerns about GO films is their susceptibility to water, aqueous media, and other polar solvents. Most of the reports on GO cross-linking are lack of information on this aspect. Actually, many of the recipes did not address this issue. In addition to that, some recipes unpreventably resulted in chemical reduction of GO due to GO's high chemical activity, while for some potential applications, reduction is not desired. At this point, GO researchers are still looking forward to solutions on those issues.

## 3.4   Doping

Due to the high catalytic activities of nitrogen(N)-containing carbon structures in several key chemical reactions, such as oxygen reduction reaction (ORR), oxygen evolution reaction (OER), and hydrogen evolution reaction (HER), N-doped GO is at the center of GO doping chemistry in literature. Several doping recipes have been introduced [49, 159–164], and more detailed discussions along with applications are offered in Chapter 3. Here we would like to show a simple example of N or S doping into GO for ORR catalysis (Fig. 3.11) [165]. The thermal reaction of GO and guest gas ($NH_3$ or $H_2S$) was exploited for the doping. In their studies, nitrogen was incorporated into graphene lattice with three types of binding configurations including pyridinic-N, pyrrolic-N, and graphitic-N, whereas sulfur was doped in a major form of thiophene-like structure. Interestingly, both N- and S-doped products showed enhanced ORR catalytic activity in alkaline media (0.1 M KOH) [165].

Like most of the reports on doping of GO or graphene, the structure characterizations of the doped atoms are usually limited with XPS analysis. Typically a high-resolution XPS scan for the targeted elements is obtained, followed by numerical deconvolution of the signals and subsequently assignments to different structures. We have to share our experience here that XPS deconvolution is not quite enough for accurate assignments of those chemical states, and other more subtle analyses are needed before any solid statement can be made. In our own experience, the binding energy position of XPS signals is easily influenced by the calibration methods used, as well as the substrate and contaminations from the ambient. Deconvolution is also quiet case dependent, easily manipulated by analysts. We would like to suggest solid-state NMR, for the dopant elements such as N, B, and S, to be a better resolve for this issue.

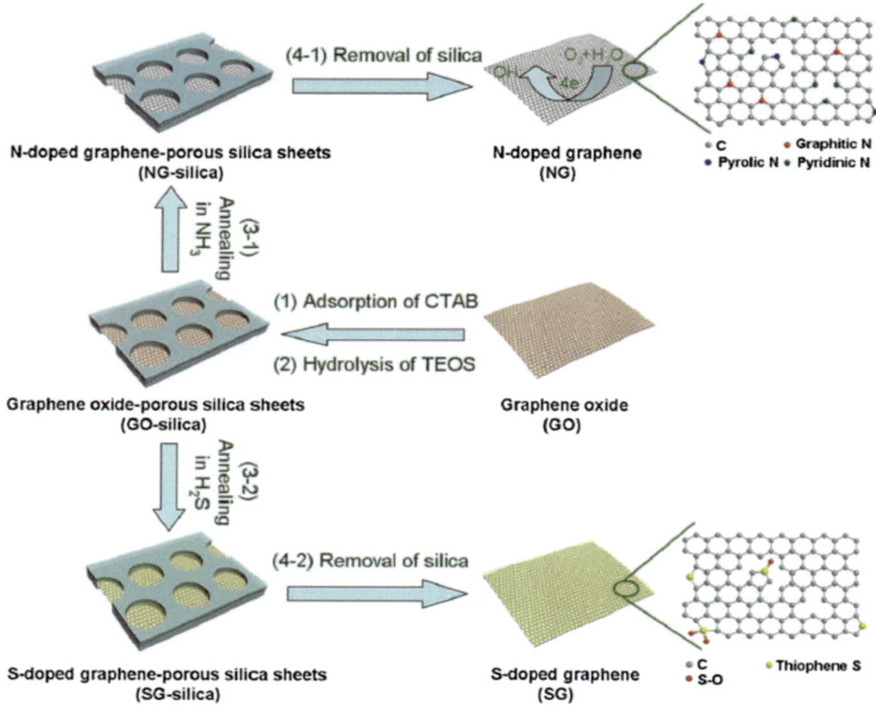



**Fig. 3.11** Schematic illustration of the fabrication of N- and S-doped graphene: (*1*, *2*) hydrolysis of TEOS around the surface of graphene oxide with the aid of a cationic surfactant, cetyltrimethyl ammonium bromide (CTAB); (*3-1*) thermal annealing of GO-silica sheets in ammonia at 600, 800, 900, and 1,000 °C, respectively; (*3-2*) thermal annealing of GO-silica sheets in $H_2S$ gas at 500, 700, and 900 °C, respectively; and (*4-1*, *4-2*) removal of silica by HF or NaOH solution. (Adapted from Ref. [165]. Copyright (2012) John Wiley & Sons Inc.)

## 3.5 Toxicity and Hygroscopicity

### 3.5.1 Toxicity

GO and RGO were reported to be toxic to *bacteria* (*Escherichia* and *Staphylococcus*) due to the cell membrane damage of the bacteria in contact with the sharp edges in GO and RGO by Akhaven et al., and hydrazine-reduced GO was more toxic than pristine GO [111]; inhibition of bacteria growth with minimal toxicity to human *alveolar epithelial A549 cells* was also shown by Hu et al. [166]. Wang et al. demonstrated that GO has dose-dependent toxicity to human fibroblast cells with obvious toxicity observed at doses above 50 µg/mL [167]. GO toxicity and blood compatibility were also reported to be dependent on dose, extent of exfoliation, and sheet size [168–170]. RGO was demonstrated to be less toxic than carbon nanotubes to PC12 cells at high concentrations by Zhang et al. [171]. PEGylated nano-sized GO only exhibited mild toxicity toward *Raji* cells at concentrations as high as 100 mg/L [172]. Paradoxically, other recent reports showed the high bio-compatibility of GO or RGO [173–177]. To address these discrepancies, Ruiz et al. reported the nonspecific enhancement of cellular growth of GO, thus verifying that GO is nontoxic to both bacteria and mammalian cells, and attributing the previous reported toxicity as effects from carry-on impurities in GO [178]. However, incorporating Ag nanoparticles onto GO matrix significantly activated its antibacterial activity [179, 180].

### 3.5.2  *Hygroscopicity*

The hydroxyl, epoxy, and carboxyl groups on GO make it very hydrophilic, and adsorbed water molecules tend to present in the interlayer voids even after prolonged drying [181]. Therefore, GO turns out to be quite hygroscopic, with water content strongly depending on the humidity of the environment [182]. When a large amount of interlamellar water is present in stacked GO films, a network of hydrogen bonds (H-bonds) forms between water molecules and those oxygenated groups on GO, thus significantly influencing its structural, mechanical, and electronic properties [181–184]. For example, as the humidity level rises, GO film swells in volume [182, 183] and tensile modulus decreases [181]. Some theoretical simulation predicted that the interlayer distance between GO flakes arises from 5.1 to 9.0 Å when water content increases from nearly zero to 26 wt%, and that as the water content exceeds 15 wt%, the H-bond network is dominated by water–water H-bonds while functional groups are indirectly connected via a chain of water molecules [185]. The dynamics of the interlamellar water has also been studied by neutron scattering, and a "two-site" jump motion mechanism has been proposed [182]. Interestingly, when water is replaced by $D_2O$, a lower interlayer distance was always observed, probably due to the lower solubility, lower reactivity, and stronger bonding of $D_2O$ compared with $H_2O$ [182]. The presence of water has also been accounted for the carbonyl formation and hole formation during GO reduction processes [186].

## 3.6  Concluding Remarks

GO is highly active chemically, mainly due to the partial coverage of its basal planes and edges with various functional groups, listed as epoxy, hydroxyl, ketone, ester, organosulfate, and lactol structures in dry state. When dispersed in polar solvents like water, it can hydrolyze to offer carboxylic acid groups or sulfate groups, which can be heavily functionalized with various chemicals as shown in literature. Activations and reactions with other moieties on GO are also well explored. The removal of these functional groups is another important topic that is inevitably related to the mass production of graphene, and hence the gold rush of graphene research in the last decade. It leads to large number of literature reports as summarized in Table 3.1. Doping and cross-linking of GO are relatively new focuses recently, and thus the investigations and explorations are in the early stage of growth.

In this chapter, we have summarized and discussed a variety of chemistry so far happened on GO, namely its reduction, functionalization, cross-linking, and doping, most of which are closely related to various applications, such as catalysis, composites, water purification, and energy conversion and storage. A detailed comparison of different reduction recipes was summarized, mainly focusing on the characterizations of a variety of reduction products. Although numerous strategies and recipes have been introduced, complete conversion of GO back to perfect graphene structure is still considered to be mission impossible. Reconstruction of highly defected rGO structures into more ordered $sp^2$ carbon grains remains to be the major obstacle, although high-temperature annealing (1,100–2,000 °C) has already been explored [187].

In addition, as it comes to applications, the high activity of GO brought problems such as selectivity and tunability of its functions. These challenges shed light on synthesis of its smaller molecular analogues, which might be more controllable if bottom-up organic synthetic strategies are explored. The blossom of GO research, partly due to its versatility in chemistry, can be successfully transferred to the oxides of other two-dimensional materials such as boron nitride (BN), graphitic carbon nitride (g-$C_3N_4$), or even small polycyclic aromatic hydrocarbons (PAHs), which are constituted of a group of materials with well-defined structures. Taking GO's ability to function as a protonic conductor [188–190] as an example, the conduction pathways and mechanisms are unclear due to the complex nature of GO structure, namely multiple functional groups that can all participate or contribute to the proton transport. The clarification of this process can be significantly simplified if investigations can be done with a well-defined PAH oxide that has only one type of functional groups present in the structure. The same scenario also works for many other applications such as catalysis and sensing.

After all, it is our hope that the analysis and comparisons in this chapter will stimulate more intensive research on GO and its related chemistry and applications. Multidisciplinary research can particularly benefit from this convergence, and we believe that substantive advances are to occur at the interfaces of chemistry, materials science, and other disciplines.

**Acknowledgement** W. G. sincerely thank for the start-up funding support from the Department of Textile Engineering, Chemistry and Science at North Carolina State University, Raleigh, NC.

# References

1. Becerril HA, Mao J, Liu Z, Stoltenberg RM, Bao Z, Chen Y (2008) Evaluation of solution-processed reduced graphene oxide films as transparent conductors. ACS Nano 2:463–470
2. Gomez-Navarro C, Weitz RT, Bittner AM, Scolari M, Mews A, Burghard M, Kern K (2007) Electronic transport properties of individual chemically reduced graphene oxide sheets. Nano Lett 7:3499–3503

3. Stankovich S, Piner RD, Chen XQ, Wu NQ, Nguyen ST, Ruoff RS (2006) Stable aqueous dispersions of graphitic nanoplatelets via the reduction of exfoliated graphite oxide in the presence of poly(sodium 4-styrenesulfonate). J Mater Chem 16:155–158
4. Si Y, Samulski ET (2008) Synthesis of water soluble graphene. Nano Lett 8:1679–1682
5. Kim J, Cote LJ, Kim F, Yuan W, Shull KR, Huang J (2010) Graphene oxide sheets at interfaces. J Am Chem Soc 132:8180–8186
6. Paredes JI, Villar-Rodil S, Martinez-Alonso A, Tascon JMD (2008) Graphene oxide dispersions in organic solvents. Langmuir 24:10560–10564
7. Cote LJ, Kim J, Tung VC, Luo JY, Kim F, Huang JX (2011) Graphene oxide as surfactant sheets. Pure Appl Chem 83:95–110
8. Cote LJ, Kim F, Huang J (2008) Langmuir-Blodgett assembly of graphite oxide single layers. J Am Chem Soc 131:1043–1049
9. Cote LJ, Kim J, Zhang Z, Sun C, Huang J (2010) Tunable assembly of graphene oxide surfactant sheets: wrinkles, overlaps and impacts on thin film properties. Soft Matter 6
10. Kim F, Cote LJ, Huang J (2009) Graphene oxide: surface activity and two-dimensional assembly. Adv Mater 22:1954–1958
11. Hofmann U, Frenzel A (1934) The reduction of graphite oxide with hydrogen sulphide. Koll Zeitschr 68:149–151
12. Kim J, Im H, Kim J-M, Kim J (2012) Thermal and electrical conductivity of Al(OH)3 covered graphene oxide nanosheet/epoxy composites. J Mater Sci 47:1418–1426
13. Stankovich S, Dikin DA, Piner RD, Kohlhaas KA, Kleinhammes A, Jia Y, Wu Y, Nguyen ST, Ruoff RS (2007) Synthesis of graphene-based nanosheets via chemical reduction of exfoliated graphite oxide. Carbon 45:1558–1565
14. Kuila T, Mishra AK, Khanra P, Kim NH, Lee JH (2013) Recent advances in the efficient reduction of graphene oxide and its application as energy storage electrode materials. Nanoscale 5:52–71
15. Luo D, Zhang G, Liu J, Sun X (2011) Evaluation criteria for reduced graphene oxide. J Phys Chem C 115:11327–11335
16. Pei S, Cheng H-M (2012) The reduction of graphene oxide. Carbon 50:3210–3228
17. Gao W, Alemany LB, Ci L, Ajayan PM (2009) New insights into the structure and reduction of graphite oxide. Nat Chem 1:403–408
18. Park S, An JH, Piner RD, Jung I, Yang DX, Velamakanni A, Nguyen ST, Ruoff RS (2008) Aqueous suspension and characterization of chemically modified graphene sheets. Chem Mater 20:6592–6594
19. Shin HJ, Kim KK, Benayad A, Yoon SM, Park HK, Jung IS, Jin MH, Jeong HK, Kim JM, Choi JY, Lee YH (2009) Efficient reduction of graphite oxide by sodium borohydride and its effect on electrical conductance. Adv Funct Mater 19:1987–1992
20. Park S, An J, Potts JR, Velamakanni A, Murali S, Ruoff RS (2011) Hydrazine-reduction of graphite-and graphene oxide. Carbon 49:3019–3023
21. Obata S, Tanaka H, Saiki K (2013) Electrical and spectroscopic investigations on the reduction mechanism of graphene oxide. Carbon 55:126–132
22. Niu Z, Chen J, Hng HH, Ma J, Chen X (2012) A leavening strategy to prepare reduced graphene oxide foams. Adv Mater 24:4144–4150
23. Tung VC, Allen MJ, Yang Y, Kaner RB (2009) High-throughput solution processing of large-scale graphene. Nat Nanotechnol 4:25–29
24. Gilje S, Han S, Wang M, Wang KL, Kaner RB (2007) A chemical route to graphene for device applications. Nano Lett 7:3394–3398
25. Stankovich S, Dikin DA, Dommett GHB, Kohlhaas KM, Zimney EJ, Stach EA, Piner RD, Nguyen ST, Ruoff RS (2006) Graphene-based composite materials. Nature 442: 282–286
26. Yun JM, Yeo JS, Kim J, Jeong HG, Kim DY, Noh YJ, Kim SS, Ku BC, Na SI (2011) Solution-processable reduced graphene oxide as a novel alternative to PEDOT:PSS hole transport layers for highly efficient and stable polymer solar cells. Adv Mater 23:4923–4928
27. Liu J, Jeong H, Liu J, Lee K, Park JY, Ahn YH, Lee S (2010) Reduction of functionalized graphite oxides by trioctylphosphine in non-polar organic solvents. Carbon 48:2282–2289

28. Moon IK, Lee J, Ruoff RS, Lee H (2010) Reduced graphene oxide by chemical graphitization. Nat Commun 1:73
29. Cataldo F, Ursini O, Angelini G (2011) Graphite oxide and graphene nanoribbons reduction with hydrogen iodide. Fuller Nanotub Car N 19:461–468
30. Das AK, Srivastav M, Layek RK, Uddin ME, Jung D, Kim NH, Lee JH (2014) Iodide-mediated room temperature reduction of graphene oxide: a rapid chemical route for the synthesis of a bifunctional electrocatalyst. J Mater Chem A 2:1332–1340
31. Esfandiar A, Akhavan O, Irajizad A (2011) Melatonin as a powerful bio-antioxidant for reduction of graphene oxide. J Mater Chem 21:10907–10914
32. Liao KH, Mittal A, Bose S, Leighton C, Mkhoyan KA, Macosko CW (2011) Aqueous only route toward graphene from graphite oxide. ACS Nano 5:1253–1258
33. Han TH, Huang Y-K, Tan ATL, Dravid VP, Huang J (2011) Steam etched porous graphene oxide network for chemical sensing. J Am Chem Soc 133:15264–15267
34. Thakur S, Karak N (2012) Green reduction of graphene oxide by aqueous phytoextracts. Carbon 50:5331–5339
35. Long Y, Zhang CC, Wang XX, Gao JP, Wang W, Liu Y (2011) Oxidation of SO(2) to SO(3) catalyzed by graphene oxide foams. J Mater Chem 21:13934–13941
36. Moon IK, Lee J, Lee H (2011) Highly qualified reduced graphene oxides: the best chemical reduction. Chem Commun 47:9681–9683
37. Jung H, Yang SJ, Kim T, Kang JH, Park CR (2013) Ultrafast room-temperature reduction of graphene oxide to graphene with excellent dispersibility by lithium naphthalenide. Carbon 63:165–174
38. Zhang S, Shao Y, Liao H, Engelhard MH, Yin G, Lin Y (2011) Polyelectrolyte-induced reduction of exfoliated graphite oxide: a facile route to synthesis of soluble graphene nanosheets. ACS Nano 5:1785–1791
39. Pei S, Zhao J, Du J, Ren W, Cheng H-M (2010) Direct reduction of graphene oxide films into highly conductive and flexible graphene films by hydrohalic acids. Carbon 48:4466–4474
40. Byon HR, Suntivich J, Shao-Horn Y (2011) Graphene-based non-noble-metal catalysts for oxygen reduction reaction in acid. Chem Mater 23:3421–3428
41. Zhou X, Zhang J, Wu H, Yang H, Zhang J, Guo S (2011) Reducing graphene oxide via hydroxylamine: a simple and efficient route to graphene. J Phys Chem C 115:11957–11961
42. Liu S, Tian J, Wang L, Sun X (2011) A method for the production of reduced graphene oxide using benzylamine as a reducing and stabilizing agent and its subsequent decoration with Ag nanoparticles for enzymeless hydrogen peroxide detection. Carbon 49:3158–3164
43. Xu LQ, Yang WJ, Neoh K-G, Kang E-T, Fu GD (2010) Dopamine-induced reduction and functionalization of graphene oxide nanosheets. Macromolecules 43:8336–8339
44. Tran DN, Kabiri S, Losic D (2014) A green approach for the reduction of graphene oxide nanosheets using non-aromatic amino acids. Carbon 76:193–202
45. Chua CK, Pumera M (2013) Selective removal of hydroxyl groups from graphene oxide. Chem Eur J 19:2005–2011
46. Wang K, Feng T, Qian M, Ding HI, Chen YW, Sun ZO (2011) The field emission of vacuum filtered graphene films reduced by microwave. Appl Surf Sci 257:5808–5812
47. Murugan AV, Muraliganth T, Manthiram A (2009) Rapid, facile microwave-solvothermal synthesis of graphene nanosheets and their polyaniline nanocomposites for energy storage. Chem Mater 21:5004–5006
48. Chen WF, Yan LF, Bangal PR (2010) Preparation of graphene by the rapid and mild thermal reduction of graphene oxide induced by microwaves. Carbon 48:1146–1152
49. Wang D-W, Wu K-H, Gentle IR, Lu GQ (2012) Anodic chlorine/nitrogen co-doping of reduced graphene oxide films at room temperature. Carbon 50:3333–3341
50. Zhou TN, Chen F, Liu K, Deng H, Zhang Q, Feng JW, Fu QA (2011) A simple and efficient method to prepare graphene by reduction of graphite oxide with sodium hydrosulfite. Nanotechnology 22:045704
51. Liao RJ, Tang ZH, Lei YD, Guo BC (2011) Polyphenol-reduced graphene oxide: mechanism and derivatization. J Phys Chem C 115:20740–20746

52. Seo M, Yoon D, Hwang KS, Kang JW, Kim J (2013) Supercritical alcohols as solvents and reducing agents for the synthesis of reduced graphene oxide. Carbon 64:207–218
53. Chen C, Chen T, Wang H, Sun G, Yang X (2011) A rapid, one-step, variable-valence metal ion assisted reduction method for graphene oxide. Nanotechnology 22(40):405602
54. Jung I, Dikin DA, Piner RD, Ruoff RS (2008) Tunable electrical conductivity of individual graphene oxide sheets reduced at "low" temperatures. Nano Lett 8:4283–4287
55. Wang S, Ang PK, Wang Z, Tang ALL, Thong JTL, Loh KP (2009) High mobility, printable, and solution-processed graphene electronics. Nano Lett 10:92–98
56. Chen WF, Yan LF (2010) Preparation of graphene by a low-temperature thermal reduction at atmosphere pressure. Nanoscale 2:559–563
57. Zhu YW, Stoller MD, Cai WW, Velamakanni A, Piner RD, Chen D, Ruoff RS (2010) Exfoliation of graphite oxide in propylene carbonate and thermal reduction of the resulting graphene oxide platelets. ACS Nano 4:1227–1233
58. Nethravathi C, Rajamathi M (2008) Chemically modified graphene sheets produced by the solvothermal reduction of colloidal dispersions of graphite oxide. Carbon 46:1994–1998
59. Zhou M, Wang YL, Zhai YM, Zhai JF, Ren W, Wang FA, Dong SJ (2009) Controlled synthesis of large-area and patterned electrochemically reduced graphene oxide films. Chem Eur J 15:6116–6120
60. Chen WF, Yan LF, Bangal PR (2010) Chemical reduction of graphene oxide to graphene by sulfur-containing compounds. J Phys Chem C 114:19885–19890
61. Gao X, Tang XS (2014) Effective reduction of graphene oxide thin films by a fluorinating agent: diethylaminosulfur trifluoride. Carbon 76:133–140
62. Ahmed MS, Han HS, Jeon S (2013) One-step chemical reduction of graphene oxide with oligothiophene for improved electrocatalytic oxygen reduction reactions. Carbon 61: 164–172
63. Ma Q, Song J, Jin C, Li Z, Liu J, Meng S, Zhao J, Guo Y (2013) A rapid and easy approach for the reduction of graphene oxide by formamidinesulfinic acid. Carbon 54:36–41
64. Fernandez-Merino MJ, Guardia L, Paredes JI, Villar-Rodil S, Solis-Fernandez P, Martinez-Alonso A, Tascon JMD (2010) Vitamin C is an ideal substitute for hydrazine in the reduction of graphene oxide suspensions. J Phys Chem C 114:6426–6432
65. Sui Z, Zhang X, Lei Y, Luo Y (2011) Easy and green synthesis of reduced graphite oxide-based hydrogels. Carbon 49:4314–4321
66. Gao J, Liu F, Liu YL, Ma N, Wang ZQ, Zhang X (2010) Environment-friendly method to produce graphene that employs vitamin C and amino acid. Chem Mater 22:2213–2218
67. Dua V, Surwade SP, Ammu S, Agnihotra SR, Jain S, Roberts KE, Park S, Ruoff RS, Manohar SK (2010) All-organic vapor sensor using inkjet-printed reduced graphene oxide. Angew Chem Int Ed 49:2154–2157
68. Chen Y, Shen Y, Sun D, Zhang H, Tian D, Zhang J, Zhu J-J (2011) Fabrication of a dispersible graphene/gold nanoclusters hybrid and its potential application in electrogenerated chemiluminescence. Chem Commun (Camb) 47:11733–5
69. Kuila T, Bose S, Khanra P, Mishra AK, Kim NH, Lee JH (2012) A green approach for the reduction of graphene oxide by wild carrot root. Carbon 50:914–921
70. Chu H-J, Lee C-Y, Tai N-H (2014) Green reduction of graphene oxide by Hibiscus sabdariffa L. to fabricate flexible graphene electrode. Carbon 80:725–733
71. Fan XB, Peng WC, Li Y, Li XY, Wang SL, Zhang GL, Zhang FB (2008) Deoxygenation of exfoliated graphite oxide under alkaline conditions: a green route to graphene preparation. Adv Mater 20:4490–4493
72. Zhu CZ, Guo SJ, Fang YX, Dong SJ (2010) Reducing sugar: new functional molecules for the green synthesis of graphene nanosheets. ACS Nano 4:2429–2437
73. Liu JB, Fu SH, Yuan B, Li YL, Deng ZX (2010) Toward a universal "adhesive nanosheet" for the assembly of multiple nanoparticles based on a protein-induced reduction/decoration of graphene oxide. J Am Chem Soc 132:7279–7281
74. Chen JL, Yan XP (2010) A dehydration and stabilizer-free approach to production of stable water dispersions of graphene nanosheets. J Mater Chem 20:4328–4332

75. Guo HL, Wang XF, Qian QY, Wang FB, Xia XH (2009) A green approach to the synthesis of graphene nanosheets. ACS Nano 3:2653–2659
76. Sundaram RS, Gomez-Navarro C, Balasubramanian K, Burghard M, Kern K (2008) Electrochemical modification of graphene. Adv Mater 20:3050–3053
77. Ping J, Wang Y, Fan K, Wu J, Ying Y (2011) Direct electrochemical reduction of graphene oxide on ionic liquid doped screen-printed electrode and its electrochemical biosensing application. Biosens Bioelectron 28:204–209
78. Guo Y, Wu B, Liu H, Ma Y, Yang Y, Zheng J, Yu G, Liu Y (2011) Electrical assembly and reduction of graphene oxide in a single solution step for use in flexible sensors. Adv Mater 23:4626–30
79. Ramesha GK, Sampath S (2009) Electrochemical reduction of oriented graphene oxide films: an in situ Raman spectroelectrochemical study. J Phys Chem C 113:7985–7989
80. Shao YY, Wang J, Engelhard M, Wang CM, Lin YH (2010) Facile and controllable electrochemical reduction of graphene oxide and its applications. J Mater Chem 20:743–748
81. Pang H, Lu Q, Gao F (2011) Graphene oxide induced growth of one-dimensional fusiform zirconia nanostructures for highly selective capture of phosphopeptides. Chem Commun (Camb) 47:11772–4
82. Ambrosi A, Pumera M (2013) Precise tuning of surface composition and electron-transfer properties of graphene oxide films through electroreduction. Chem Eur J 19:4748–4753
83. Cui J, Lai Y, Wang W, Li H, Ma X, Zhan J (2014) Galvanic displacement induced reduction of graphene oxide. Carbon 66:738–741
84. Fan Z, Wang K, Wei T, Yan J, Song L, Shao B (2011) An environmentally friendly and efficient route for the reduction of graphene oxide by aluminum powder. Carbon 48:1686–1689
85. Mei X, Ouyang J (2011) Ultrasonication-assisted ultrafast reduction of graphene oxide by zinc powder at room temperature. Carbon 49:5389–5397
86. Wang G, Yang J, Park J, Gou X, Wang B, Liu H, Yao J (2008) Facile synthesis and characterization of graphene nanosheets. J Phys Chem C 112:8192–8195
87. Wu ZS, Ren WC, Gao LB, Liu BL, Jiang CB, Cheng HM (2009) Synthesis of high-quality graphene with a pre-determined number of layers. Carbon 47:493–499
88. Li CC, Yu H, Yan Q, Hng HH (2015) Green synthesis of highly reduced graphene oxide by compressed hydrogen gas towards energy storage devices. J Power Sources 274:310–317
89. Matsumoto Y, Koinuma M, Ida S, Hayami S, Taniguchi T, Hatakeyama K, Tateishi H, Watanabe Y, Amano S (2011) Photoreaction of graphene oxide nanosheets in water. J Phys Chem C 115:19280–19286
90. Williams G, Seger B, Kamat PV (2008) $TiO_2$-graphene nanocomposites. UV-Assisted photocatalytic reduction of graphene oxide. ACS Nano 2:1487–1491
91. Ji T, Hua Y, Sun M, Ma N (2013) The mechanism of the reaction of graphite oxide to reduced graphene oxide under ultraviolet irradiation. Carbon 54:412–418
92. Prezioso S, Perrozzi F, Donarelli M, Stagnini E, Treossi E, Palermo V, Santucci S, Nardone M, Moras P, Ottaviano L (2014) Dose and wavelength dependent study of graphene oxide photoreduction with VUV synchrotron radiation. Carbon 79:478–485
93. Cote LJ, Cruz-Silva R, Huang JX (2009) Flash reduction and patterning of graphite oxide and its polymer composite. J Am Chem Soc 131:11027–11032
94. Le Borgne V, Bazi H, Hayashi T, Kim YA, Endo M, El Khakani MA (2014) Hydrogen-assisted pulsed KrF-laser irradiation for the *in situ* photoreduction of graphene oxide films. Carbon 77:857–867
95. Trusovas R, Ratautas K, Račiukaitis G, Barkauskas J, Stankevičienė I, Niaura G, Mažeikienė R (2013) Reduction of graphite oxide to graphene with laser irradiation. Carbon 52:574–582
96. Huang L, Liu Y, Ji L-C, Xie Y-Q, Wang T, Shi W-Z (2011) Pulsed laser assisted reduction of graphene oxide. Carbon 49:2431–2436
97. Dumée LF, Feng C, He L, Yi Z, She F, Peng Z, Gao W, Banos C, Davies JB, Huynh C (2014) Single step preparation of meso-porous and reduced graphene oxide by gamma-ray irradiation in gaseous phase. Carbon 70:313–318

98. Zhang Y, Li D, Tan X, Zhang B, Ruan X, Liu H, Pan C, Liao L, Zhai T, Bando Y (2013) High quality graphene sheets from graphene oxide by hot-pressing. Carbon 54:143–148
99. Chang DW, Choi H-J, Jeon I-Y, Seo J-M, Dai L, Baek J-B (2014) Solvent-free mechanochemical reduction of graphene oxide. Carbon 77:501–507
100. Eigler S, Dotzer C, Hirsch A (2012) Visualization of defect densities in reduced graphene oxide. Carbon 50:3666–3672
101. Berger C, Song Z, Li X, Wu X, Brown N, Naud C, Mayou D, Li T, Hass J, Marchenkov AN, Conrad EH, First PN, de Heer WA (2006) Electronic confinement and coherence in patterned epitaxial graphene. Science 312:1191–1196
102. Behera SK (2011) Enhanced rate performance and cyclic stability of Fe(3)O(4)-graphene nanocomposites for Li ion battery anodes. Chem Commun 47:10371–10373
103. Boukhvalov DW, Katsnelson MI (2008) Modeling of graphite oxide. J Am Chem Soc 130:10697–10701
104. Gao X, Jang J, Nagase S (2009) Hydrazine and thermal reduction of graphene oxide: reaction mechanisms, product structures, and reaction design. J Phys Chem C 114:832–842
105. Kim MC, Hwang GS, Ruoff RS (2009) Epoxide reduction with hydrazine on graphene: a first principles study. J Chem Phys 131:064704
106. Dreyer DR, Todd AD, Bielawski CW (2014) Harnessing the chemistry of graphene oxide. Chem Soc Rev 43:5288–5301
107. Tang X-Z, Li W, Yu Z-Z, Rafiee MA, Rafiee J, Yavari F, Koratkar N (2011) Enhanced thermal stability in graphene oxide covalently functionalized with 2-amino-4, 6-didodecylamino-1, 3, 5-triazine. Carbon 49:1258–1265
108. Liu ZB, Xu YF, Zhang XY, Zhang XL, Chen YS, Tian JG (2009) Porphyrin and fullerene covalently functionalized graphene hybrid materials with large nonlinear optical properties. J Phys Chem B 113:9681–9686
109. Niyogi S, Bekyarova E, Itkis ME, McWilliams JL, Hamon MA, Haddon RC (2006) Solution properties of graphite and graphene. J Am Chem Soc 128:7720–7721
110. Xu YF, Liu ZB, Zhang XL, Wang Y, Tian JG, Huang Y, Ma YF, Zhang XY, Chen YS (2009) A graphene hybrid material covalently functionalized with porphyrin: synthesis and optical limiting property. Adv Mater 21:1275–1279
111. Akhavan O, Ghaderi E (2010) Toxicity of graphene and graphene oxide nanowalls against bacteria. ACS Nano 4:5731–5736
112. Zhang YJ, Hu WB, Li B, Peng C, Fan CH, Huang Q (2011) Synthesis of polymer-protected graphene by solvent-assisted thermal reduction process. Nanotechnology 22
113. Zhu J, Li Y, Chen Y, Wang J, Zhang B, Zhang J, Blau WJ (2011) Graphene oxide covalently functionalized with zinc phthalocyanine for broadband optical limiting. Carbon 49:1900–1905
114. Hu X, Mu L, Wen J, Zhou Q (2012) Covalently synthesized graphene oxide-aptamer nanosheets for efficient visible-light photocatalysis of nucleic acids and proteins of viruses. Carbon 50:2772–2781
115. Mejias Carpio IE, Mangadlao JD, Nguyen HN, Advincula RC, Rodrigues DF (2014) Graphene oxide functionalized with ethylenediamine triacetic acid for heavy metal adsorption and anti-microbial applications. Carbon 77:289–301
116. Wu H, Shi H, Wang Y, Jia X, Tang C, Zhang J, Yang S (2014) Hyaluronic acid conjugated graphene oxide for targeted drug delivery. Carbon 69:379–389
117. Li Z-F, Zhang H, Liu Q, Liu Y, Stanciu L, Xie J (2014) Covalently-grafted polyaniline on graphene oxide sheets for high performance electrochemical supercapacitors. Carbon 71:257–267
118. Veca LM, Lu FS, Meziari MJ, Cao L, Zhang PY, Qi G, Qu LW, Shrestha M, Sun YP (2009) Polymer functionalization and solubilization of carbon nanosheets. Chem Commun 2565–2567
119. Yang YF, Wang J, Zhang J, Liu JC, Yang XL, Zhao HY (2009) Exfoliated graphite oxide decorated by PDMAEMA chains and polymer particles. Langmuir 25:11808–11814
120. Wan Y-J, Tang L-C, Gong L-X, Yan D, Li Y-B, Wu L-B, Jiang J-X, Lai G-Q (2014) Grafting of epoxy chains onto graphene oxide for epoxy composites with improved mechanical and thermal properties. Carbon 69:467–480

121. Dreyer DR, Jarvis KA, Ferreira PJ, Bielawski CW (2011) Graphite oxide as a dehydrative polymerization catalyst: a one-step synthesis of carbon-reinforced poly(phenylene methylene) composites. Macromolecules 44:7659–7667
122. Eda G, Mattevi C, Yamaguchi H, Kim H, Chhowalla M (2009) Insulator to semimetal transition in graphene oxide. J Phys Chem C 113:15768–15771
123. Liu J, Chen G, Jiang M (2011) Supramolecular hybrid hydrogels from noncovalently functionalized graphene with block copolymers. Macromolecules 44:7682–7691
124. Krueger M, Berg S, Stone DA, Strelcov E, Dikin DA, Kim J, Cote LJ, Huang J, Kolmakov A (2011) Drop-casted self-assembling graphene oxide membranes for scanning electron microscopy on wet and dense gaseous samples. ACS Nano 5:10047–10054
125. Hu HT, Wang XB, Wang JC, Liu FM, Zhang M, Xu CH (2011) Microwave-assisted covalent modification of graphene nanosheets with chitosan and its electrorheological characteristics. Appl Surf Sci 257:2637–2642
126. Park S, Dikin DA, Nguyen ST, Ruoff RS (2009) Graphene oxide sheets chemically cross-linked by polyallylamine. J Phys Chem C 113:15801–15804
127. An Z, Compton OC, Putz KW, Brinson LC, Nguyen ST (2011) Bio-inspired borate cross-linking in ultra-stiff graphene oxide thin films. Adv Mater 23:3842–3846
128. Gonçalves G, Marques PAAP, Barros-Timmons A, Bdkin I, Singh MK, Emami N, Grácio J (2010) Graphene oxide modified with PMMA *via* ATRP as a reinforcement filler. J Mater Chem 20:9927–9934
129. McGrail BT, Rodier BJ, Pentzer E (2014) Rapid functionalization of graphene oxide in water. Chem Mater 26:5806–5811
130. Eigler S, Hu Y, Ishii Y, Hirsch A (2013) Controlled functionalization of graphene oxide with sodium azide. Nanoscale 5:12136–12139
131. Kamada S, Nomoto H, Fukuda K, Fukawa T, Shirai H, Kimura M (2011) Noncovalent wrapping of chemically modified graphene with pi-conjugated disk-like molecules. Colloid Polym Sci 289:925–932
132. Lomeda JR, Doyle CD, Kosynkin DV, Hwang WF, Tour JM (2008) Diazonium functionalization of surfactant-wrapped chemically converted graphene sheets. J Am Chem Soc 130:16201–16206
133. Gao W, Majumder M, Alemany LB, Narayanan TN, Ibarra MA, Pradhan BK, Ajayan PM (2011) Engineered graphite oxide materials for application in water purification. Acs Appl Mater Inter 3:1821–1826
134. Ballesteros-Garrido R, Rodriguez R, Álvaro M, Garcia H (2014) Photochemistry of covalently functionalized graphene oxide with phenothiazinyl units. Carbon 74:113–119
135. Liu F, Chung S, Oh G, Seo TS (2011) Three-dimensional graphene oxide nanostructure for fast and efficient water-soluble dye removal. Acs Appl Mater Inter 4:922–927
136. Liang J, Huang Y, Zhang L, Wang Y, Ma Y, Guo T, Chen Y (2009) Molecular-level dispersion of graphene into poly (vinyl alcohol) and effective reinforcement of their nanocomposites. Adv Funct Mater 19:2297–2302
137. Hu K, Gupta MK, Kulkarni DD, Tsukruk VV (2013) Ultra-robust graphene oxide-silk fibroin nanocomposite membranes. Adv Mater 25:2301–2307
138. An SJ, Zhu YW, Lee SH, Stoller MD, Emilsson T, Park S, Velamakanni A, An JH, Ruoff RS (2010) Thin film fabrication and simultaneous anodic reduction of deposited graphene oxide platelets by electrophoretic deposition. J Phys Chem Lett 1:1259–1263
139. Layek RK, Das AK, Park MJ, Kim NH, Lee JH (2015) Enhancement of physical, mechanical, and gas barrier properties in noncovalently functionalized graphene oxide/poly (vinylidene fluoride) composites. Carbon 81:329–338
140. Bai H, Xu Y, Zhao L, Li C, Shi G (2009) Non-covalent functionalization of graphene sheets by sulfonated polyaniline. Chem Commun 1667–1669
141. Teng C-C, Ma C-CM LC-H, Yang S-Y, Lee S-H, Hsiao M-C, Yen M-Y, Chiou K-C, Lee T-M (2011) Thermal conductivity and structure of non-covalent functionalized graphene/epoxy composites. Carbon 49:5107–5116

142. Paci JT, Belytschko T, Schatz GC (2007) Computational studies of the structure, behavior upon heating, and mechanical properties of graphite oxide. J Phys Chem C 111:18099–18111
143. Satti A, Larpent P, Gun'ko Y (2010) Improvement of mechanical properties of graphene oxide/poly (allylamine) composites by chemical crosslinking. Carbon 48:3376–3381
144. Li YQ, Yu T, Yang TY, Zheng LX, Liao K (2012) Bio-inspired nacre-like composite films based on graphene with superior mechanical, electrical, and biocompatible properties. Adv Mater 24:3426–3431
145. Cheng Q, Wu M, Li M, Jiang L, Tang Z (2013) Ultratough artificial nacre based on conjugated cross-linked graphene oxide. Angew Chem Int Ed 125:3838–3843
146. Tian Y, Cao Y, Wang Y, Yang W, Feng J (2013) Realizing ultrahigh modulus and high strength of macroscopic graphene oxide papers through crosslinking of mussel-inspired polymers. Adv Mater 25:2980–2983
147. Hu H, Brown PH, Labavitch JM (1996) Species variability in boron requirement is correlated with cell wall pectin. J Exp Bot 47:227–232
148. Sudeep PM, Narayanan TN, Ganesan A, Shaijumon MM, Yang H, Ozden S, Patra PK, Pasquali M, Vajtai R, Ganguli S, Roy AK, Anantharaman MR, Ajayan PM (2013) Covalently interconnected three-dimensional graphene oxide solids. ACS Nano 7:7034–7040
149. Wei W, Yang S, Zhou H, Lieberwirth I, Feng X, Müllen K (2013) 3D graphene foams cross-linked with pre-encapsulated $Fe_3O_4$ nanospheres for enhanced lithium storage. Adv Mater 25:2909–2914
150. Wan W, Li L, Zhao Z, Hu H, Hao X, Winkler DA, Xi L, Hughes TC, Qiu J (2014) Ultrafast fabrication of covalently cross-linked multifunctional graphene oxide monoliths. Adv. Funct, Mater
151. Sun H, Xu Z, Gao C (2013) Synergistically assembled carbon aerogels. Adv Mater 25:2554–2560
152. Huang H, Chen P, Zhang X, Lu Y, Zhan W (2013) Edge-to-edge assembled graphene oxide aerogels with outstanding mechanical performance and superhigh chemical activity. Small 9:1397–1404
153. Hu H, Zhao Z, Wan W, Gogotsi Y, Qiu J (2013) Ultralight and highly compressible graphene aerogels. Adv Mater 25:2219–2223
154. Yang X, Qiu L, Cheng C, Wu Y, Ma ZF, Li D (2011) Ordered gelation of chemically converted graphene for next-generation electroconductive hydrogel films. Angew Chem Int Ed 50:7325–7328
155. Cong HP, Wang P, Yu SH (2014) Highly elastic and superstretchable graphene oxide/polyacrylamide hydrogels. Small 10:448–453
156. Wu L, Ohtani M, Takata M, Saeki A, Seki S, Ishida Y, Aida T (2014) Magnetically induced anisotropic orientation of graphene oxide locked by in situ hydrogelation. ACS Nano 8(5):4640–9
157. Compton OC, Cranford SW, Putz KW, An Z, Brinson LC, Buehler MJ, Nguyen ST (2012) Tuning the mechanical properties of graphene oxide paper and its associated polymer nanocomposites by controlling cooperative intersheet hydrogen bonding. ACS Nano 6:2008–2019
158. Arndt KF (2006) Hydrogel sensors and actuators. Frontiers 4
159. Zhang H, Kuila T, Kim NH, Yu DS, Lee JH (2014) Simultaneous reduction, exfoliation, and nitrogen doping of graphene oxide via a hydrothermal reaction for energy storage electrode materials. Carbon 69:66–78
160. Van Khai T, Na HG, Kwak DS, Kwon YJ, Ham H, Shim KB, Kim HW (2012) Influence of N-doping on the structural and photoluminescence properties of graphene oxide films. Carbon 50:3799–3806
161. Li M, Wu Z, Ren W, Cheng H, Tang N, Wu W, Zhong W, Du Y (2012) The doping of reduced graphene oxide with nitrogen and its effect on the quenching of the material's photoluminescence. Carbon 50:5286–5291
162. Liu Y, Feng Q, Tang N, Wan X, Liu F, Lv L, Du Y (2013) Increased magnetization of reduced graphene oxide by nitrogen-doping. Carbon 60:549–551

163. Liu Y, Feng Q, Xu Q, Li M, Tang N, Du Y (2013) Synthesis and photoluminescence of F and N co-doped reduced graphene oxide. Carbon 61:436–440
164. Yang J, Jo MR, Kang M, Huh YS, Jung H, Kang Y-M (2014) Rapid and controllable synthesis of nitrogen doped reduced graphene oxide using microwave-assisted hydrothermal reaction for high power-density supercapacitors. Carbon 73:106–113
165. Yang S, Zhi L, Tang K, Feng X, Maier J, Müllen K (2012) Efficient synthesis of heteroatom (N or S)-doped graphene based on ultrathin graphene oxide-porous silica sheets for oxygen reduction reactions. Adv Funct Mater 22:3634–3640
166. Chen S, Chen P, Wang Y (2011) Carbon nanotubes grown in situ on graphene nanosheets as superior anodes for Li-ion batteries. Nanoscale 3:4323–4329
167. Wang K, Ruan J, Song H, Zhang JL, Wo Y, Guo SW, Cui DX (2013) Biocompatibility of graphene oxide. Nanoscale Res Lett 8(1):393
168. Liao KH, Lin YS, Macosko CW, Haynes CL (2011) Cytotoxicity of graphene oxide and graphene in human erythrocytes and skin fibroblasts. Acs Appl Mater Inter 3:2607–2615
169. Wojtoniszak M, Chen X, Kalenczuk RJ, Wajda A, Lapczuk J, Kurzewski M, Drozdzik M, Chu PK, Borowiak-Palen E (2012) Synthesis, dispersion, and cytocompatibility of graphene oxide and reduced graphene oxide. Colloids Surf B Biointerfaces 89:79–85
170. Begurn P, Ikhtiari R, Fugetsu B (2011) Graphene phytotoxicity in the seedling stage of cabbage, tomato, red spinach, and lettuce. Carbon 49:3907–3919
171. Pan DY, Wang S, Zhao B, Wu MH, Zhang HJ, Wang Y, Jiao Z (2009) Li storage properties of disordered graphene nanosheets. Chem Mater 21:3136–3142
172. Sun XM, Liu Z, Welsher K, Robinson JT, Goodwin A, Zaric S, Dai HJ (2008) Nano-graphene oxide for cellular imaging and drug delivery. Nano Res 1:203–212
173. Agarwal S, Zhou XZ, Ye F, He QY, Chen GCK, Soo J, Boey F, Zhang H, Chen P (2010) Interfacing live cells with nanocarbon substrates. Langmuir 26:2244–2247, http://pubs.rsc.org/en/content/articlehtml/2010/jm/c0jm01674h
174. Chang YL, Yang ST, Liu JH, Dong E, Wang YW, Cao AN, Liu YF, Wang HF (2011) In vitro toxicity evaluation of graphene oxide on A549 cells. Toxicol Lett 200:201–210
175. Ryoo SR, Kim YK, Kim MH, Min DH (2010) Behaviors of NIH-3 T3 fibroblasts on graphene/carbon nanotubes: proliferation, focal adhesion, and gene transfection studies. ACS Nano 4:6587–6598
176. Park S, Mohanty N, Suk JW, Nagaraja A, An JH, Piner RD, Cai WW, Dreyer DR, Berry V, Ruoff RS (2010) Biocompatible, robust free-standing paper composed of a TWEEN/graphene composite. Adv Mater 22:1736–40
177. Chen H, Muller MB, Gilmore KJ, Wallace GG, Li D (2008) Mechanically strong, electrically conductive, and biocompatible graphene paper. Adv Mater 20:3557–61
178. Ruiz ON, Fernando KAS, Wang BJ, Brown NA, Luo PG, McNamara ND, Vangsness M, Sun YP, Bunker CE (2011) Graphene oxide: a nonspecific enhancer of cellular growth. ACS Nano 5:8100–8107
179. Bao Q, Zhang D, Qi P (2011) Synthesis and characterization of silver nanoparticle and graphene oxide nanosheet composites as a bactericidal agent for water disinfection. J Colloid Interface Sci 360:463–470
180. Das MR, Sarma RK, Saikia R, Kale VS, Shelke MV, Sengupta P (2011) Synthesis of silver nanoparticles in an aqueous suspension of graphene oxide sheets and its antimicrobial activity. Coll Surf B Biointerf 83:16–22
181. Dikin DA, Stankovich S, Zimney EJ, Piner RD, Dommett GHB, Evmenenko G, Nguyen ST, Ruoff RS (2007) Preparation and characterization of graphene oxide paper. Nature 448:457–460
182. Buchsteiner A, Lerf A, Pieper J (2006) Water dynamics in graphite oxide investigated with neutron scattering. J Phys Chem B 110:22328–22338
183. Lerf A, Buchsteiner A, Pieper J, Sch枚ttl S, Dekany I, Szabo T, Boehm HP (2006) Hydration behavior and dynamics of water molecules in graphite oxide. J Phys Chem Solid 67:1106–1110

184. Jung I, Dikin D, Park S, Cai W, Mielke SL, Ruoff RS (2008) Effect of water vapor on electrical properties of individual reduced graphene oxide sheets. J Phys Chem C 112:20264–20268
185. Medhekar NV, Ramasubramaniam A, Ruoff RS, Shenoy VB (2010) Hydrogen bond networks in graphene oxide composite paper: structure and mechanical properties. ACS Nano 4:2300–2306
186. Acik M, Mattevi C, Gong C, Lee G, Cho K, Chhowalla M, Chabal YJ (2010) The role of intercalated water in multilayered graphene oxide. ACS Nano 4:5861–5868
187. Song L, Khoerunnisa F, Gao W, Dou W, Hayashi T, Kaneko K, Endo M, Ajayan PM (2013) Effect of high-temperature thermal treatment on the structure and adsorption properties of reduced graphene oxide. Carbon 52:608–612
188. Gao W, Singh N, Song L, Liu Z, Reddy ALM, Ci LJ, Vajtai R, Zhang Q, Wei BQ, Ajayan PM (2011) Direct laser writing of micro-supercapacitors on hydrated graphite oxide films. Nat Nanotechnol 6:496–500
189. Gao W, Wu G, Janicke MT, Cullen DA, Mukundan R, Baldwin JK, Brosha EL, Galande C, Ajayan PM, More KL (2014) Ozonated graphene oxide film as a proton-exchange membrane. Angew Chem Int Ed 53:3588–3593
190. Karim MR, Hatakeyama K, Matsui T, Takehira H, Taniguchi T, Koinuma M, Matsumoto Y, Akutagawa T, Nakamura T, Noro S-I (2013) Graphene oxide nanosheet with high proton conductivity. J Am Chem Soc 135:8097–8100

# Chapter 4
# GO/rGO as Advanced Materials for Energy Storage and Conversion

Gang Wu and Wei Gao

**Abstract** Recently, GO/rGO has become promising platforms for advanced materials in energy related technologies. Extensive applications of rGO have been explored in a variety of electrochemical energy storage and conversion technologies (e.g., fuel cells, metal–air batteries, supercapacitors, and water splitting devices). In particular, GO/rGO was studied as efficient components in catalysts in fuel cells and metal–air batteries for the oxygen reduction reaction (ORR) and oxygen evolution reaction (OER), one pair of the most important electrochemical reactions. The promising applications are primarily due to their unique physical and chemical properties, such as high surface area, access to large quantities, tunable electronic/ionic conductivity, unique graphitic basal plane structure, and the easiness of modification or functionalization. Chemical doping with heteroatoms (e.g., N, B, P, or S) into graphitic domains can tune the electronic properties, provide more active sites, and enhance the interaction between carbon structures and oxygen molecules. Meanwhile, GO/rGO has demonstrated excellent performances in electric double-layer capacitors (EDLCs, also known as supercapacitors or ultracapacitors) with excellent volumetric and gravimetric capacitance densities as well as feasibility in design and fabrications. Additionally, rGO holds great promise to be high-performance anode materials in lithium-ion batteries due to its favorable interactions with Li. In this chapter, we will discuss the uses of GO/rGO derivatives in these energy conversion and storage technologies, providing insights and guidance for further optimization and design of multifunctional materials for energy applications.

**Keywords** Graphene oxide • Electrocatalysis • Fuel cells • Supercapacitors • Lithium batteries

G. Wu (✉)
Department of Chemical and Biological Engineering, University at Buffalo,
The State University of New York, Buffalo, NY 14260, USA
e-mail: gangwu@buffalo.edu

W. Gao
The Department of Textile Engineering, Chemistry & Science, College of Textiles,
North Carolina State University, Raleigh, NC 27695, USA
e-mail: wgao5@ncsu.edu

© Springer International Publishing Switzerland 2015
W. Gao (ed.), *Graphene Oxide*, DOI 10.1007/978-3-319-15500-5_4

## 4.1  GO/rGO-Derived Nonprecious Metal Catalysts

### 4.1.1  Overview

Energy conversion and storage via direct electrochemical reactions are one of the most important energy technologies of modern day, including polymer electrolyte fuel cells (PEFCs), metal (Li or Zn)-air batteries, and water electrolyzers. These devices offer many advantages over traditional fossil fuel combustion technologies, including improved overall efficiency, high energy density, and the significant reduction of $CO_2$ and other emissions. PEFCs are high-efficiency chemical-to-electrical energy conversion devices that can be used as power sources in electric vehicles and portable and stationary applications. Metal–air batteries can provide vastly enhanced energy densities over traditional lithium-ion batteries. For example, compared to other rechargeable systems, $Li–O_2$ batteries, as a possible next-generation energy storage and conversion device, have attracted much attention due to their very high energy density. Unlike the traditional intercalation electrodes used in Li-ion batteries, the porous cathode in the $Li–O_2$ cell is capable of taking reactant $O_2$ from the atmosphere, instead of storing bulky reactants in the electrode. As a result, the battery has significantly improved specific energy density (theoretical value of 5,200 Wh/kg) [1], which has great promise to meet the battery targets set for automotive applications (1,700 Wh/kg, derived from the practical energy density of gasoline) [2, 3]. Additionally, these metal–air batteries can be widely used to store clean energy obtained from renewable wind and solar sources and can also be used for grid-scale energy storage in power plants. Noteworthy, both fuel cells and metal–air cells have similar oxygen cathodes and share the same operating principles for the ORR and OER [4].

Due to the high overpotentials of ORR and OER and inherently slow reaction kinetics, the performance of these electrochemical energy technologies is greatly limited by ORR and OER activities [5–7]. Pt- and Ir-based catalysts represent the state of the art for the ORR and OER, respectively, in terms of their highest activity and durability. To realize a sustainable development, replacing these precious metals with highly active and stable nonprecious metal catalysts (NPMCs) has thereby become a very important topic in the field of electrocatalysis. Importantly, exploring advanced catalyst designs and synthesis strategies using earth-abundant elements (e.g., Fe, Co, N, and C) will be a scientific interest with the potential to provide fundamental knowledge and understanding in the field of chemical catalysis and electrochemical science.

Graphene oxide (GO) is a major precursor for preparing graphene and has attracted substantial attention recently. GO is usually made through oxidation of graphite powder under harsh chemical conditions. The most effective method currently used is Hummers method involving a few strong oxidants (e.g., $KMnO_4$, $H_2SO_4$, and $K_2S_2O_8$). As a result, different oxygen-containing groups, especially hydroxyl, epoxy, and carboxyl, are formed on both sides and at the edges. Importantly, GO flakes can be exfoliated into single-layer and multilayer sheets

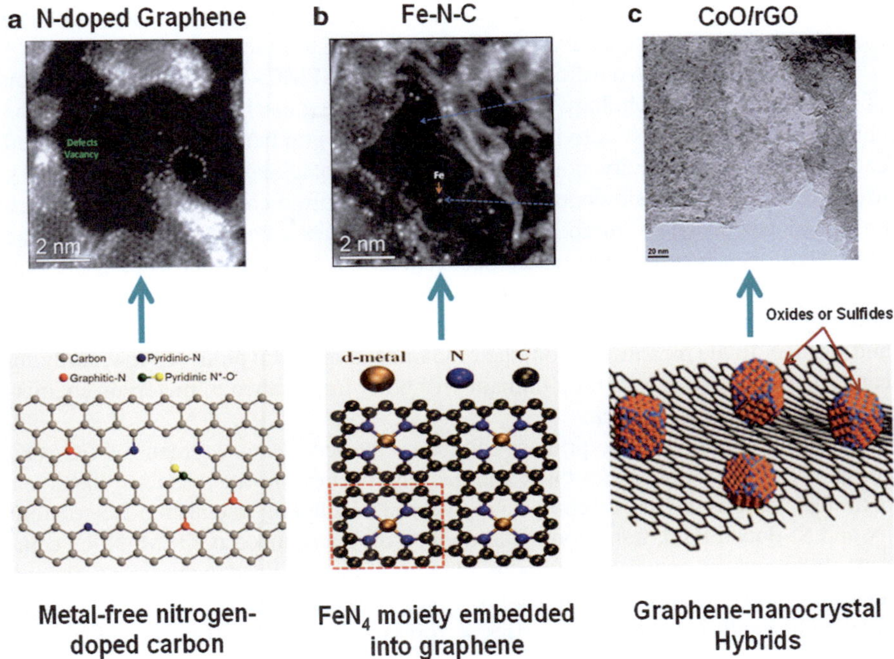



**Fig. 4.1** Currently studied graphene-based catalysts: (**a**) metal-free nitrogen-doped graphene. Reprinted with permission from Ref. [9]. Copyright 2012 American Chemical Society (**b**) transition metal incorporated graphene. Reproduced from Ref [8] by permission of the Royal Society of Chemistry. (**c**) Transition metal-compound/graphene hybrid catalysts. Reproduced from Ref [10] by permission of the Royal Society of Chemistry

when dissolved in water due to the strong coulombic repulsion between hydrolyzed sheets. In order to prepare graphene materials, a complete reduction process for GO is needed to remove all of oxygen-containing functionalities from GO surface to prepare electrically conductive reduced GO (rGO).

Generally, the current applications of GO/rGO in ORR catalysts can be divided into three categories (Fig. 4.1) [8–10]. (1) At first, heteroatoms, such as N, can dope into graphitic basal planes and change the electronic and structural properties, providing more active sites and enhancing the interaction between carbon structure and oxygen molecules. Theoretical studies have shown that doped nitrogen atoms can be viewed as $n$-type carbon dopants that disorder carbon lattices and donate electrons to carbon, thus facilitating the ORR activity. (2) Secondly, there is increasing evidence showing transition metal cations, especially Fe, are able to bond with nitrogen and embedded into the basal planes. This configuration was proposed as a possible active site with much improved intrinsic activity, compared to metal-free nitrogen-doped graphene structures. (3) In the third case, transition metal nanocrystals including oxides and sulfides have been found to be catalytically active for ORR or OER. When depositing such nanocrystal onto active $sp^2$ carbon lattice, a synergetic effect may rise between the active species in graphene planes and these loaded transition metal nanocrystals, thereby leading to much improved activity.

In our recent effort to develop high-performance NPMCs for ORR, the formation of graphene-like morphology in catalysts is more closely associated with the enhancement of catalyst activity and stability. The graphene-rich $TiO_2$-supported catalyst has higher activity, compared to the Ketjenblack-supported catalysts [11]. Meanwhile, the graphene-containing MWNTs supported catalysts exhibited much enhanced performance durability, compared to other catalyst without graphene [12]. Thus, the existence of graphene seems to have a promotional role in improving catalyst performance. These results suggest that GO/rGO-based catalysts provide new opportunities on catalyst development for electrochemical energy conversion and storage. In the meantime, from the fundamental research point of view, studying catalytic properties of graphitic domain will be of importance to materials chemistry and engineering research.

Thus, in this chapter, we primarily focus on the recent development of GO/rGO composite electrocatalysts for ORR and their applications in fuel cells and metal–air batteries. Various GO/rGO composite materials including metal-free heteroatom (N and S)-doped rGO, transition metal-derived nitrogen-doped rGO (M–N–C) catalysts, and transition metal-compound/rGO hybrids are discussed in terms of their correlations among synthesis, structures/morphology, ORR activity, and the corresponding alkaline fuel cells or metal–air batteries performance under real working conditions.

## *4.1.2 Heteroatom-Doped rGO Catalysts*

Recently, the research on exploring the use of graphene nanocomposites as metal-free catalysts has been one of the major subjects for the ORR cathode for alkaline fuel cells and metal–air batteries [13]. Due to the low cost, environmental acceptability, corrosion resistance, high electrical conductivity, and good ORR activity in alkaline media, graphene materials are viewed as ideal candidates for metal-free catalyst for these energy storage and conversion technologies. Generally, among different allotropes of carbon, graphene with ordered graphitic structure are expected to facilitate the electron transfer rate and exhibit good electrochemical stability. Importantly, it has been commonly recognized that heteroatom doping into $sp^2$ carbon lattice leads to much improved ORR activity [14, 15]. In this section, we discuss the synthesis, structures, and applications of such metal-free heteroatom-doped rGO composite catalysts.

### 4.1.2.1 Nitrogen-Doped rGO Catalysts

While traditional carbon materials (e.g., carbon blacks, nanotube, fibers) are capable of being good catalyst supports, they generally exhibit insufficient catalytic activity for the ORR [16]. Historically, several methods have been used to modify

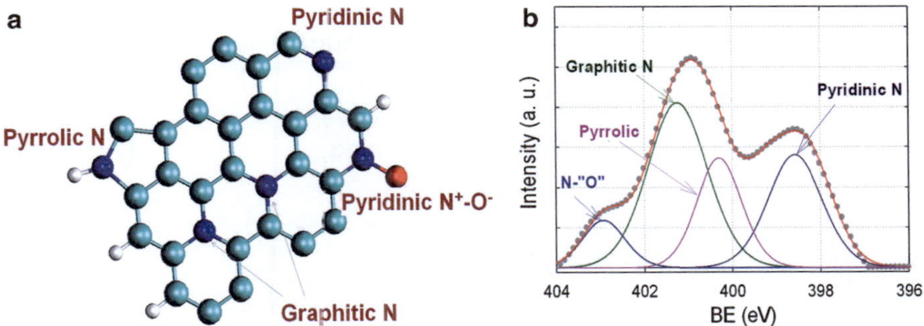


**Fig. 4.2** (**a**) Nitrogen doping into carbon plane at different location; (**b**) typical binding energies of nitrogen atoms in different doping environment determined by X-ray photoelectron spectroscopy. Reproduced from Ref [17] by permission of the Royal Society of Chemistry

these carbon materials in an effort to improve their catalytic performance. Apart from optimization of morphological properties (e.g., surface area and porosity), chemical modification of the carbon has been explored to improve intrinsically catalytic activity. Importantly, chemical doping with heteroatoms (e.g., N, B, P, or S) into carbon lattice can tune the electronic and geometric properties, provide more active sites, and enhance the interaction between carbon structure and oxygen molecules. Among various heteroatoms, nitrogen doping plays a critical role in modifying carbon structures due to comparable atomic size of nitrogen and carbon, as well as the presence of five valence electrons in the nitrogen atoms available to form strong covalence bonds with carbon atoms.

As shown in Fig. 4.2 [17], various nitrogen doping into carbon can be identified by different binding energies in X-ray photoelectron spectroscopy (XPS) associated with dominant pyridinic (398.6±0.3 eV) and graphitic nitrogen (401.3±0.3 eV) [18]. Pyridinic nitrogen is obtained by doping at the edge of the graphene layer, contributing one $p\pi$ electron to the graphitic $\pi$ system. Quaternary nitrogen is the result of in-plane doping and contributes two $p\pi$ electrons. In addition, the pyrrolic form of nitrogen (400.5±0.3 eV) observed at the relatively low heat-treatment temperature (600–700 °C) is assigned to nitrogen atoms in a pentagon structure, which is indistinguishable in XPS from pyridone (pyridinic nitrogen next to an OH group) [19, 20]. Both pyrrolic and pyridone nitrogens have been shown to decompose at temperatures above 800 °C to either pyridinic or quaternary nitrogen [19–22]. Nitrogen species with a high binding energy (402–405 eV) can be assigned to oxidized nitrogen, such as a pyridinic $N^+$–$O^-$ species [19]. In particular, quaternary and pyridinic nitrogens are dominant nitrogen doping in such nitrogen-doped carbon catalysts and crucial to ORR activity.

In the case of graphitic nitrogen, before the doping, density functional theory (DFT) calculations indicated that the electron density of carbon planes is uniform showing insignificant activity for ORR. However, doping of graphitic N into graphene structure leads to a nonuniform electron distribution, especially when two quaternary N are doped into the same hexagon unit. As a result, the nonuniform

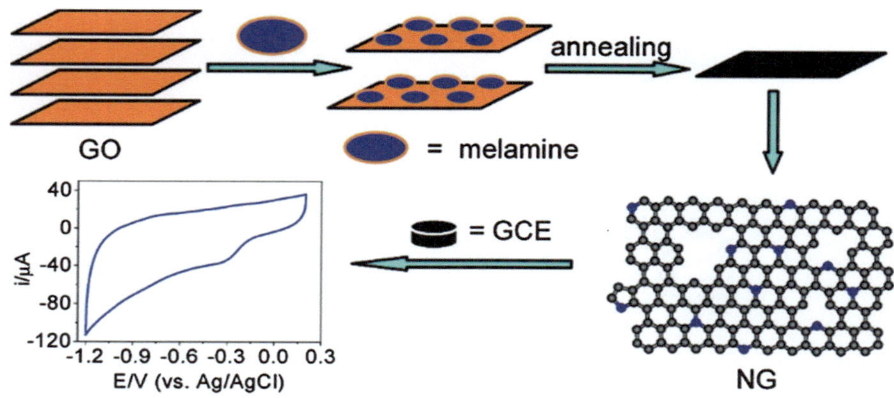



**Fig. 4.3** Illustration of the nitrogen doping process of melamine into GO layers [1]. Melamine adsorbed on the surfaces of GO when temperature is <300 °C [2]. Melamine condensed and formed carbon nitride when temperature is <600 °C [3]. Carbon nitride decomposed and doped into graphene layers when temperature is >600 °C. Reprinted with permission from Ref [32]. Copyright (2011) American Chemical Society

electron density which resulted from graphitic N doping will enhance catalytic activity of carbon. In the meantime, from the geometric point of view [23], doping of graphitic nitrogen in plane will make the C–N bonds much shorter and is comparable to O–O bonds. This change will facilitate $O_2$ adsorption and subsequent disassociation of O–O bonds. The pyridinic N sites have been widely received as catalytic active sites for ORR, because XPS experiments showed that the highly catalytic carbon materials usually have a large amount of pyridinic nitrogen. Using graphene nano-ribbon models, the DFT calculation also indicated the pyridinic N doped at the edge can reduce the energy barrier for oxygen adsorption on adjacent carbon atoms. In addition, doping of N at the edge of graphene can accelerate the rate-limiting first-electron transfer during the ORR [24].

Currently, nitrogen functionalization of rGO can be realized by several methods, such as heat annealing with $NH_3$ [25, 26], arc discharge [27, 28], nitrogen plasma [29, 30], and hydrothermal reaction with $NH_4OH$ [31]. As a result, a great number of N-doped rGO electrocatalysts have been developed toward the ORR. Recently, a facile, catalyst-free thermal annealing approach was proposed for large-scale synthesis of N-doped rGO using low-cost industrial material melamine as the nitrogen source (Fig. 4.3) [32]. Typically, GO powder and melamine were ground together in a mortar using pestle for about 5 min, and the mixture was heat-treated at designed reaction temperatures (e.g., 800 °C) for 1 h at a flow of argon [32]. This approach can completely avoid the contamination of transition metal catalysts, and thus, the intrinsic catalytic performance of metal-free N-doped graphene can be investigated. The atomic percentage of nitrogen in the doped graphene samples can be tuned up to 10.1 at.%. It was found that dominant nitrogen atoms are in the form of pyridine, doped at the edge of carbon planes. The resulting N–rGO exhibited excellent electrocatalytic activity for the ORR in alkaline electrolytes, similar to what was

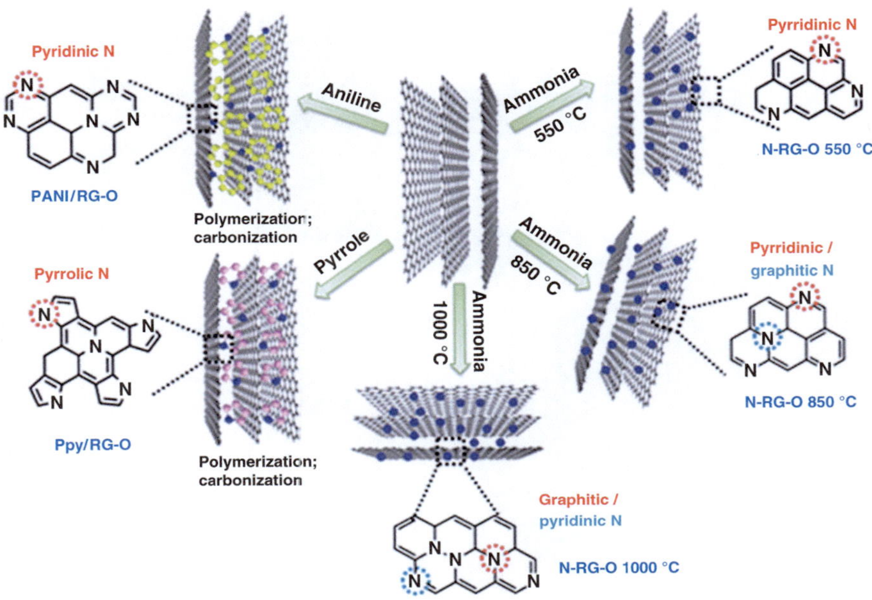


**Fig. 4.4** Schematic diagram for the preparation of N-doped graphene with different N states. N–rGO 550, 850, and 1,000 °C are prepared by annealing of GO powder at temperatures of 550, 850, and 1,000 °C under a $NH_3 \cdot N$ precursor. PANi/rGO and PPy/rGO are prepared by annealing of PANII/GO and PPy/GO composites at 850 °C. Reproduced from Ref [33] by permission of the Royal Society of Chemistry

observed in nitrogen-doped vertically aligned carbon nanotubes prepared by CVD or other methods [14, 15]. Importantly, the measured electrocatalytic activity with the N–rGO is not greatly dependent on nitrogen content in rGO. It may suggest that, besides of nitrogen doping, morphology and structure of rGO is also important for the electrocatalytic activity toward the ORR.

In order to understand the relationship between the type and concentration of N species and their corresponding catalytic activity, a series of N-doped rGO catalysts were prepared by annealing GO under ammonia or different N-containing polymers (e.g., polyaniline—PANI, polypyrrole—PPy) [33], as demonstrated in Fig. 4.4. The observed superior ORR catalytic activity of N–rGO catalysts prepared by annealing of GO with various N-containing polymer composites is directly linked to both pyridinic N and graphitic N. Annealing of PANI/rGO and PPy/rGO leads to predominate pyridinic N and pyrrolic N species, respectively. Most importantly, the electrocatalytic activity of N-containing metal-free catalysts is highly dependent on the graphitic N content while pyridinic N species improve the onset potential for ORR. In good agreement with others, the total atomic content of N in the metal-free, graphene-based catalyst did not play an important role in the ORR process. Graphitic N can greatly increase the limiting current density, while pyridinic N species might convert the ORR reaction mechanism from a $2e^-$-dominated process to a $4e^-$-dominated process [33].

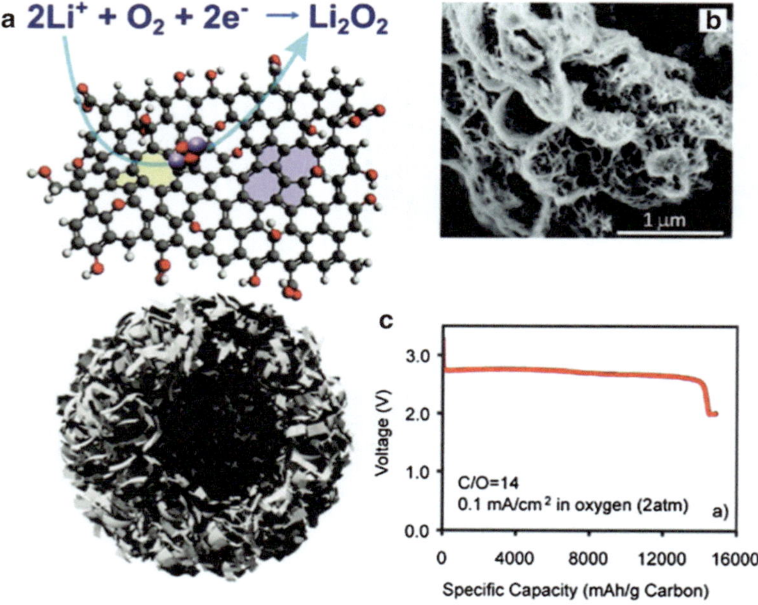



**Fig. 4.5** (**a**) Schematic structure of a functionalized graphene sheet (*upper image*) with an ideal bimodal porous structure (*lower image*) which is highly desirable for Li–O$_2$ battery operation. (**b**) SEM images of as-prepared rGO (C/O = 14). (**c**) The discharge curve of a Li–O$_2$ cell using rGO(C/O = 14) as the air electrode ($P_{O2}$ = 2 atm). Reprinted with permission from Ref. [35]. Copyright (2011) American Chemical Society

rGO also has become one of the most promising catalysts for the ORR in nonaqueous electrolyte for Li–O$_2$ batteries. Recently, rGO nanosheets (GNSs) have been successfully established in an air electrode for nonaqueous batteries [31, 34]. The air electrode based on rGO GNSs yielded a high discharge capacity of ~8,700 mAh g$^{-1}$ at a current density of 75 mA g$^{-1}$, compared to ~1,900 mAh g$^{-1}$ for BP-2000 and ~1,050 mAh g$^{-1}$ for Vulcan XC-72, respectively. The dominant discharge product was Li$_2$CO$_3$ and a small amount of Li$_2$O$_2$. This result indicated that rGO can be used as an advanced cathode catalyst for Li–air batteries. Recently, Xiao and coworkers used a colloidal microemulsion approach and demonstrated the construction of hierarchically porous air electrodes with functionalized rGO sheets that contain lattice defects and hydroxyl, epoxy, and carboxyl groups. Figure 4.5a shows the schematic structure of functionalized graphene with an ideal bimodal porous structure which is highly desirable for Li–O$_2$ battery operation. Unlike conventional two-dimensional (2D) layered graphene sheets that hinder the rapid gas diffusion that is essential for the efficient operation of Li–air batteries, the three-dimensional (3D) air electrodes developed in this work via the thermal expansion and simultaneous reduction of GO consisting of interconnected pore channels on both the micro- and nanometer length scales (Fig. 4.5b) are ideally suited for air electrodes since the pores on different length scales may facilitate oxygen diffusion, avoiding pore blockage due to $Li_2O_2$ deposition during discharge in nonaqueous $Li–O_2$ batteries. DFT calculation and electron microscopy characterization suggest that the functional groups and lattice defects play a critical role in improving the battery performance by forming small particle size of discharge products. In particular, the $C/O = 14$ graphene electrode, containing more functional groups, resulting in better performance and smaller $Li_2O_2$ particles, when compared to those of $C/O = 100$ with less functional groups. Importantly, in the unique hierarchical structures, large tunnels constructed by the macropores facilitate oxygen transfer into the air electrode, and small "pores" are able to provide ideal multiphase regions for the ORR. Thus, the combination of the abundance of functional groups and the hierarchical pore structures leads to an exceptionally high capacity of 15,000 mAh $g^{-1}$ during the discharge process (Fig. 4.5c).

### 4.1.2.2 Other Heteroatom-Doped rGO Catalysts

In addition to nitrogen doping into carbon materials, other elements such as boron, phosphorous, and sulfur are also able to dope into carbon lattice, modifying geometric and electronic structures of graphitic planes and affecting the catalytic activity toward the ORR.

Sulfur (electronegativity of sulfur: 2.58) has a close electronegativity to carbon (electronegativity of carbon: 2.55). S doping into $sp^2$ carbon lattice may play the similar role to nitrogen in enhancing catalytic activity. It was found by Bandosz and coworkers that rGO in confined space of silica gel nanopores doped with sulfur shows high catalytic activity for the ORR in alkaline medium and exhibits a superior tolerance to the presence of methanol [36]. In particular, the treatment with hydrogen sulfide at 800 °C of rGO resulted in an incorporation of 2.8 at.% sulfur in the rGO layers. XPS analysis indicated that this sulfur is mainly in the form of elemental sulfur/polysulfides (0.18 at.%), R–SH groups (2.28 at.%), C–S–C/R–S$_2$–OR configuration (0.21 at.%), and $R_2–S=O$ (0.13 at.%) functionalities. The good performance of the S-doped material is linked to the coexistence of sulfur and oxygen on the surface in equal atomic quantities and a unique porosity being the replica of the silica pores. The former leads to the positive charge on the carbon atoms, which are the reaction sites. Importantly, enhanced hydrophobicity of the surface and the resulting meso/micropores could further enhance the adsorption of $O_2$. In addition, N and S dual-doped mesoporous graphene has also been reported as a metal-free catalyst for ORR in alkaline medium [37]. The DFT calculations have revealed that the introduction of sulfur to the carbon matrix can be associated with the mismatch of the outermost orbitals of sulfur and carbon due to their close electronegativity. Thus, the S atom is positively charged and hence can be viewed as the catalytic center for ORR. The synergistic performance enhancement is due to the redistribution of spin and charge densities resulting from the dual doping of S and N atoms, which leads to a large number of active carbon atom sites.

In addition, Yang et al., demonstrated a successful preparation of sulfur-doped graphene (S-graphene) by directly annealing GO and benzyl disulfide (BDS) in

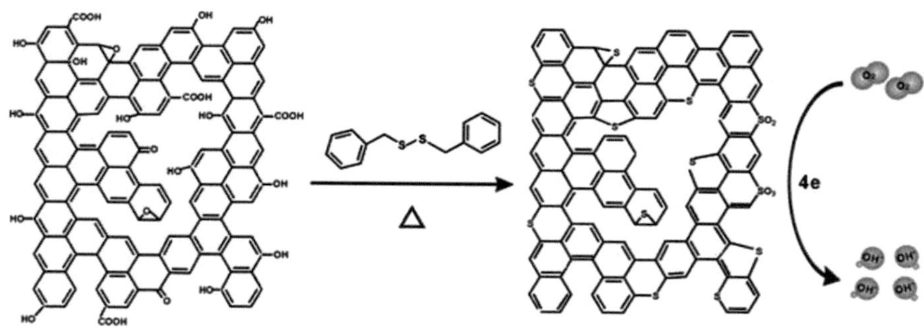

**Fig. 4.6** Schematic illustration of S-graphene preparation. Reprinted with permission from Ref. [38]. Copyright (2011) American Chemical Society

argon (Fig. 4.6) [38]. Typically, the experimental scheme for S-graphene preparation is illustrated in a typical procedure—GO and BDS were first ultrasonically dispersed in ethanol. The resulting suspension was spread onto an evaporating dish and dried at 40 °C, forming a uniform solid mixture. The mixture was placed in a quartz tube under an argon atmosphere and annealed at 600–1,050 °C. The final products were collected from the quartz tube. The contents and bonding configurations of sulfur in these S-graphenes can be adjusted by varying the mass ratios of GO and BDS or the annealing temperature. The prepared graphene sheets are wrinkled and folded. The S distribution in the plane of S-doped graphene is relatively uniform. The electrocatalytic activity shows that the S-graphene can exhibit excellent catalytic activity, long-term stability, and high methanol tolerance in alkaline media for ORRs. Moreover, we find also that the graphene doped with another element which has a similar electronegativity as carbon such as selenium (electronegativity of selenium: 2.55) shows a similarly high ORR catalytic activity. The experimental results are believed to be significant because they not only give further insight into the ORR mechanism of these metal-free doped carbon materials but also open a way to fabricate other new low-cost NPMCs with high electrocatalytic activity by a simple, economical, and scalable approach for fuel cell applications.

DFT calculations revealed that electron-deficient boron dopants were positively charged in the B-doped carbon lattice and induced the chemisorptions of oxygen molecules on doped carbon. The electrocatalytic activity derived from B doping for the ORR is derived from the electron accumulation in the vacant $2pz$ orbital of boron dopant from the $\pi^*$ electrons of the conjugated system, which then transfers to the chemisorbed oxygen molecules through boron as a bridge. The transferred charge is able to weaken the O–O bonds and thus facilitate the ORR [39].

Recently, Xia et al. [40] reported a facile, effective, and catalyst-free thermal annealing approach to preparing boron-doped graphene (Fig. 4.7) on a large scale using boron oxide as the boron source. Boron-doped graphene was synthesized through thermal annealing GO, synthesized by the modified Hummers method in the presence of boron oxide ($B_2O_3$), where boron atoms coming from $B_2O_3$ vapor can replace carbon atoms within graphene structures at high temperature in a homemade tubular furnace. In a typical procedure, GO powder was put onto the surface of $B_2O_3$

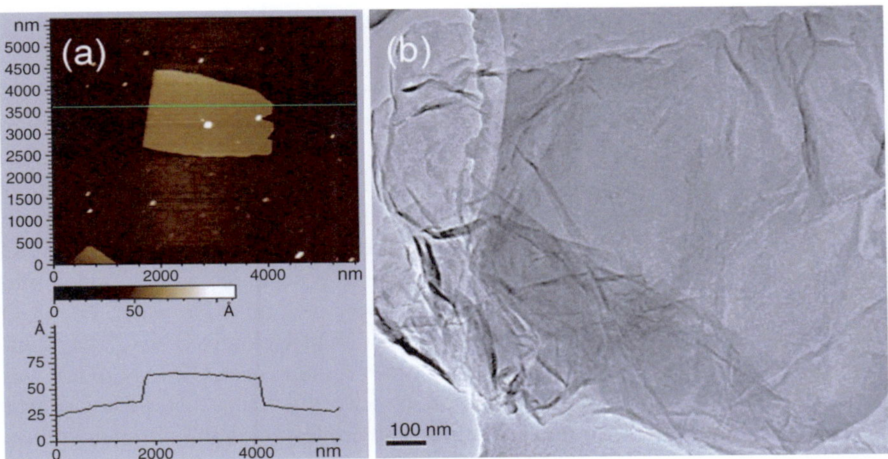



**Fig. 4.7** Typical AFM image (**a**) and low-resolution TEM image (**b**) of BG. Reproduced from Ref [40] by permission of the Royal Society of Chemistry

in a corundum crucible, which was then placed in the center of a corundum tube with a continuous flow of argon to guarantee an inert atmosphere in the tube furnace. The center temperature of the furnace was heated to 1,200 °C at an increasing rate of 5 °C min$^{-1}$. After maintaining at this temperature for 4 h, the sample was cooled to room temperature slowly under an Ar atmosphere. The obtained product was then refluxed in 3 M NaOH aqueous solution for 2 h to remove any of the unreacted boron oxide. After filtration and water washing, the product was dried in a vacuum at 60 °C. For comparison, pristine graphene was also prepared using a similar procedure but without adding $B_2O_3$. Compared to pristine graphene, the product BG exhibits excellent electrocatalytic activity and long-term stability toward ORR in alkaline electrolytes and good tolerance to poisons. Although the functioning mechanism of doped boron in graphene is not clear yet, this will not prevent us using it as a promising metal-free catalyst toward ORR in fuel cells and other catalytic applications.

In summary, above are discussion of heteroatom-doped metal-free rGO catalysts in terms of their synthesis, structures/morphologies, and the resulting catalytic activity for the ORR in various electrolytes for electrochemical energy storage and conversion. It is scientifically important to explore the enhancement of catalytic activity of rGO by modifying its electronic and geometric structure through chemical dopants. It is worth noting that the currently studied metal-free rGO catalysts are capable of efficiently catalyzing the ORR in alkaline media, even exceeding the Pt catalysts. However, these metal-free rGO catalysts always suffer the low activity and high peroxide yield during the ORR in acidic electrolyte. The variance of ORR activity in both alkaline and acidic media suggests that, among different possible active sites (CN$x$, MN$x$ or MCN$x$, M=Co or Fe), the most active site structures for the ORR in alkaline and acidic electrolytes are likely not the same, due to their different reaction mechanisms. Especially, the abundant CN$_x$ structures in alkaline media may be active enough to catalyze ORR, but still insufficient in acidic media.

### 4.1.3   Transition Metal-Doped rGO Catalysts

As we discussed above, although the metal-free rGO catalysts exhibited good catalytic activity in alkaline media, they are currently suffering poor activity in more challenging acid electrolyte that is very important for Nafion®-based polymer electrolyte fuel cells. However, it was found that the incorporation of transition metals especially Fe can lead to significant enhancement of the ORR activity of rGO in acidic electrolyte. Yang Shao-Horn' group first reported such type of Fe-based rGO catalysts [41]. As shown in Fig. 4.8, the synthesis procedure can be divided into three steps. At first, g-C$_3$N$_4$ and FeCl$_3$ (6 wt%) were well mixed in water. Then GO was added to the suspension followed by a reduction using a mixed NH$_3$/N$_2$H$_4$ solution at 80–100 °C. Secondly, the resulting powder mixture (FeCl$_3$/g-C$_3$N$_4$/rGO) was collected after drying via vacuum filtration. Noteworthy, the g-C$_3$N$_4$ precursor was polymerized thermally from dicyandiamide and employed as the nitrogen source. The resulting FeCl$_3$/g-C$_3$N$_4$/rGO mixture showed a homogeneous dark green color that was derived from the yellowish FeCl$_3$ and g-C$_3$N$_4$ and the black rGO.

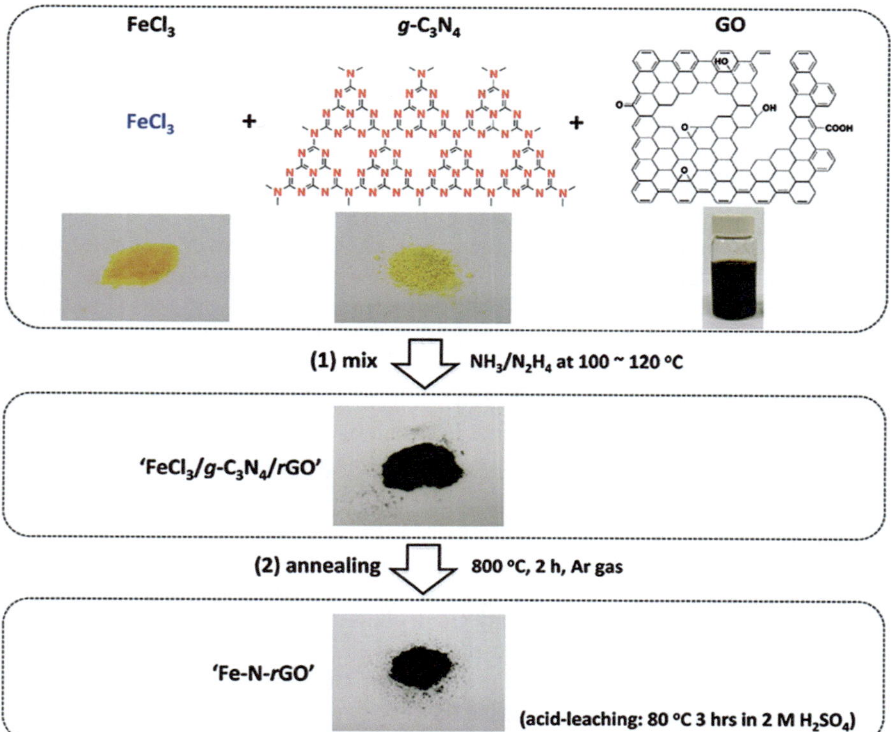


**Fig. 4.8** Schematic illustration of Fe–N–rGO synthesis and optical images of powder samples at various stages of synthesis. Reprinted with permission from Ref. [41]. Copyright (2013) American Chemical Society

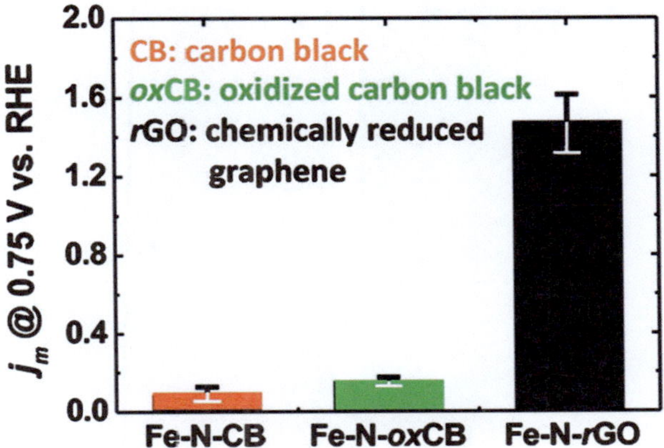


**Fig. 4.9** Comparison of ORR mass activity from RDE measurements in $O_2$-saturated 0.5 M $H_2SO_4$ at 10 mV s$^{-1}$ of scan rate and 900 rpm of rotation rate. Reprinted with permission from Ref. [41]. Copyright (2013) American Chemical Society

In the third step, the mixture was heat-treated at 800 °C for 2 h under Ar atmosphere. During the high temperature treatment, significant mass loss was observed, especially at 200–300 °C for the removal of residual oxygen-containing groups on rGO. Additionally, the second mass loss of ~65 wt% occurs in the temperature range from 620 to 700 °C due to sublimation of g-$C_3N_4$.

As shown in Fig. 4.9, the ORR mass activity of the Fe–N–rGO was determined at 0.75 V vs. RHE, yielding a current density of ~1.5 mA mg$^{-1}$. This value is seven times higher compared to other carbon black or oxidized carbon black supported Fe–N–C catalysts. The stability of Fe–N–rGO for ORR was further evaluated at a constant potential at 0.5 V in $O_2$-bubbled 0.5 M $H_2SO_4$ at 80 °C. Fe–N–rGO was found to have higher stability than other traditional carbon black supported Fe–N–C catalysts. It is hypothesized that the enhanced stability of Fe–N–rGO for ORR is due to enhanced degree of graphitization of Fe–N–rGO. In addition, significantly reduced $H_2O_2$ during the ORR observed with the Fe–N–rGO catalysts is beneficial for improving the stability. Thus, the rGO-based Fe catalyst exhibited much enhanced ORR mass activity and stability approaching those of the state-of-the-art non-noble-metal catalysts reported to date. This effort demonstrated that utilizing the unique surface chemistry of rGO is able to create active Fe–$N_x$ sites and develop highly active and stable catalysts for the ORR in PEFC.

### 4.1.4   rGO/Oxide Hybrids Catalysts

Metal oxides, especially manganese and cobalt oxides, are found for a long time to be active for oxygen reaction in alkaline solution [42, 43]. However, inherent low conductivity is one of the important drawbacks that limit their application as ORR

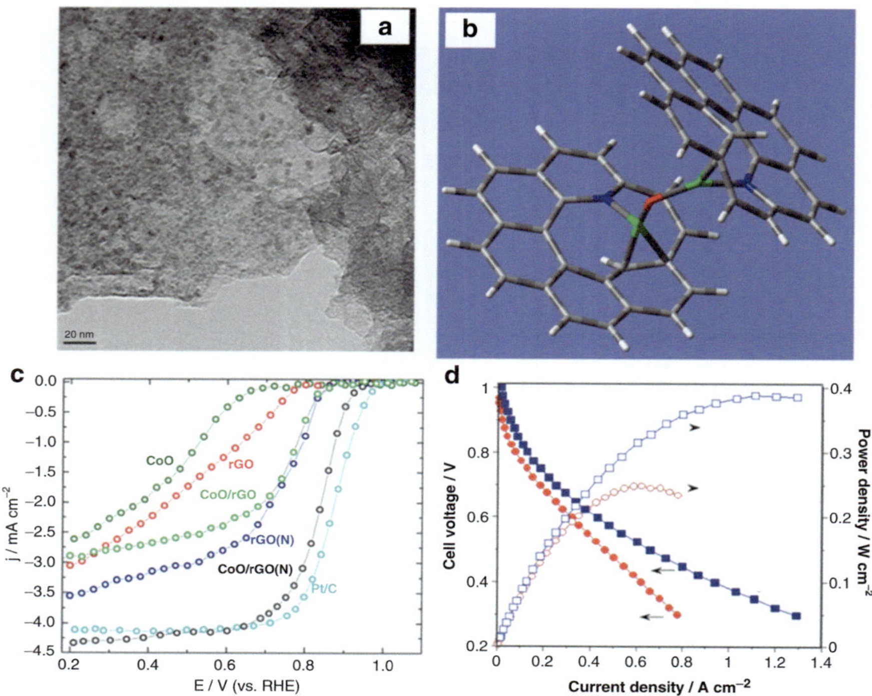

**Fig. 4.10** (**a**) CoO/rGO(N) catalyst morphology. (**b**) Optimized rGO(N)–Co(II)–O–Co(II)–rGO(N) using highest occupied molecular orbital (HOMO) at the B3LYP/6-31G level of theory. Carbon (*gray*), nitrogen (*blue*), cobalt (*green*), and oxygen (*red*). (**c**) Steady-state RDE polarization plots for ORR on CoO/rGO(N) catalyst and other controls (CoO, rGO, rGO(N), Pt/C) in 0.1 M $O_2$-saturated KOH at 25 °C and 900 rpm. (**d**) Anion-exchange-membrane $H_2$ (at 1 atm, 57 % RH)/$O_2$ (at 1 atm, 100 % RH) fuel cell tests using Pt/C (*square symbol*) and CoO/rGO(N) cathodes (*circle symbol*) at 60 °C. Reprinted with permission from Ref. [48]. Copyright (2013) American Chemical Society

cathodes in alkaline fuel cell. In order to overcome this challenge, these simple metal oxides need to be combined with other highly electrically conductive materials for preparing composite ORR electrocatalysts. Recently, a class of nanocarbon/transition metal oxides or sulfides hybrid catalysts for ORR in alkaline media were extensively studied in Dai's group at Stanford University by rGO, which includes $Co_3O_4$/N–rGO [31], $MnCo_2O_4$/N–rGO [42], $Co_{1-x}S$/rGO [44], etc. Besides manganese- or cobalt-based catalysts, a large variety of metal oxides have also been studied as ORR catalyst in alkaline solution, including $Cu_2O$/rGO [45, 46] and $Fe_3O_4$/N–rGO aerogel [47].

Recently, we reported a novel graphene composite CoO/rGO(N) catalyst containing high loading cobalt oxide (24.7 wt%, Co) via depositing CoO nanoparticles onto nitrogen-doped rGO (rGO(N)), as shown in (Fig. 4.10a) [48]. The Co(II) is identified as a dominant cobalt species on catalysts and most likely coordinately coupled with pyridinic N doped into graphene planes, evidenced by X-ray absorption spectra and DFT calculations (Fig. 4.10b). In particular, DFT calculations suggest that CoO

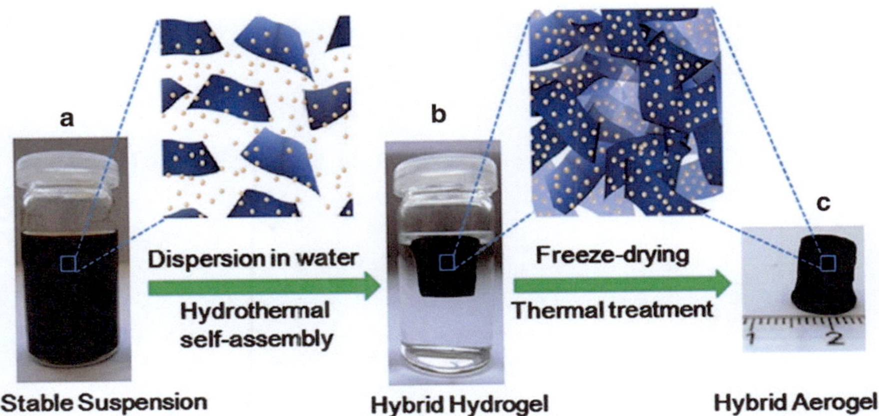



**Fig. 4.11** Fabrication process for the 3D $Fe_3O_4$/N–rGO catalyst. (**a**) Stable suspension of GO, iron ions, and PPy dispersed in a vial. (**b**) Fe- and PPy-supporting graphene hybrid hydrogel prepared by hydrothermal self-assembly and floating on water in a vial and its ideal assembled model. (**c**) Monolithic $Fe_3O_4$/N–rGO hybrid aerogel obtained after freeze-drying and thermal treatment. Reprinted with permission from Ref. [47]. Copyright (2012) American Chemical Society

can strongly couple with pyridinic nitrogens in the rGO(N) and thus form a stable structure represented as rGO(N)–Co(II)–O–Co(II)–rGO(N). With this unique structure, a synergistic effect between rGO(N) and cobalt oxide may facilitate the ORR in alkaline media, yielding a much improved activity ($E_{1/2}$ ~0.83 V vs. RHE) and four electron selectivity, when compared to either rGO(N) or CoO along (Fig. 4.10c). The developed catalysts were also tested in anion-exchange-membrane alkaline fuel cells using the AS4 anion exchange ionomer (Tokuyama). Importantly, due to the richness of defects and nitrogen doping, the graphene-based supports can accommodate a high Co loading, leading to a thin cathode layer with enhanced mass transfer in alkaline fuel cell cathode. As a result, the thickness of the catalyst layer of the MEAs is less than 20 nm, which is close to the MEA prepared with Pt/C catalysts. Compared to the Pt/C cathode, the cell with the CoO/rGO(N) cathode only has a slightly decreased OCV around 38 mV (Fig. 4.10d). A similar downward voltage shift at the low current density range was observed for the CoO/rGO(N) and Pt/C cathodes. At typical operating voltage of ~0.6 V, the power outcome of the CoO/rGO(N) catalyst is approaching that of the Pt/C catalyst.

Three-dimensional (3D) N-doped rGO aerogel-supported $Fe_3O_4$ nanoparticles ($Fe_3O_4$/N–rGO) as efficient cathode catalysts for the oxygen reduction reaction (ORR) are reported (Fig. 4.11) [47]. As displayed in Fig. 4.12 [47], the $Fe_3O_4$/N–rGO hybrids exhibit an interconnected macroporous framework of graphene sheets with uniform dispersion of $Fe_3O_4$ nanoparticles. XRD patter versifies the formation of $Fe_3O_4$ during the synthesis. SEM image indicates a significant portion of the nanoparticles which are encapsulated within the graphene layers, suggesting efficient assembly between the $Fe_3O_4$ and the rGO sheets. In studying the effects of the carbon support on the $Fe_3O_4$ for the ORR, $Fe_3O_4$/N–GAs show a more positive onset potential, higher oxygen reduction density, lower $H_2O_2$ yield, and higher electron transfer number in alkaline

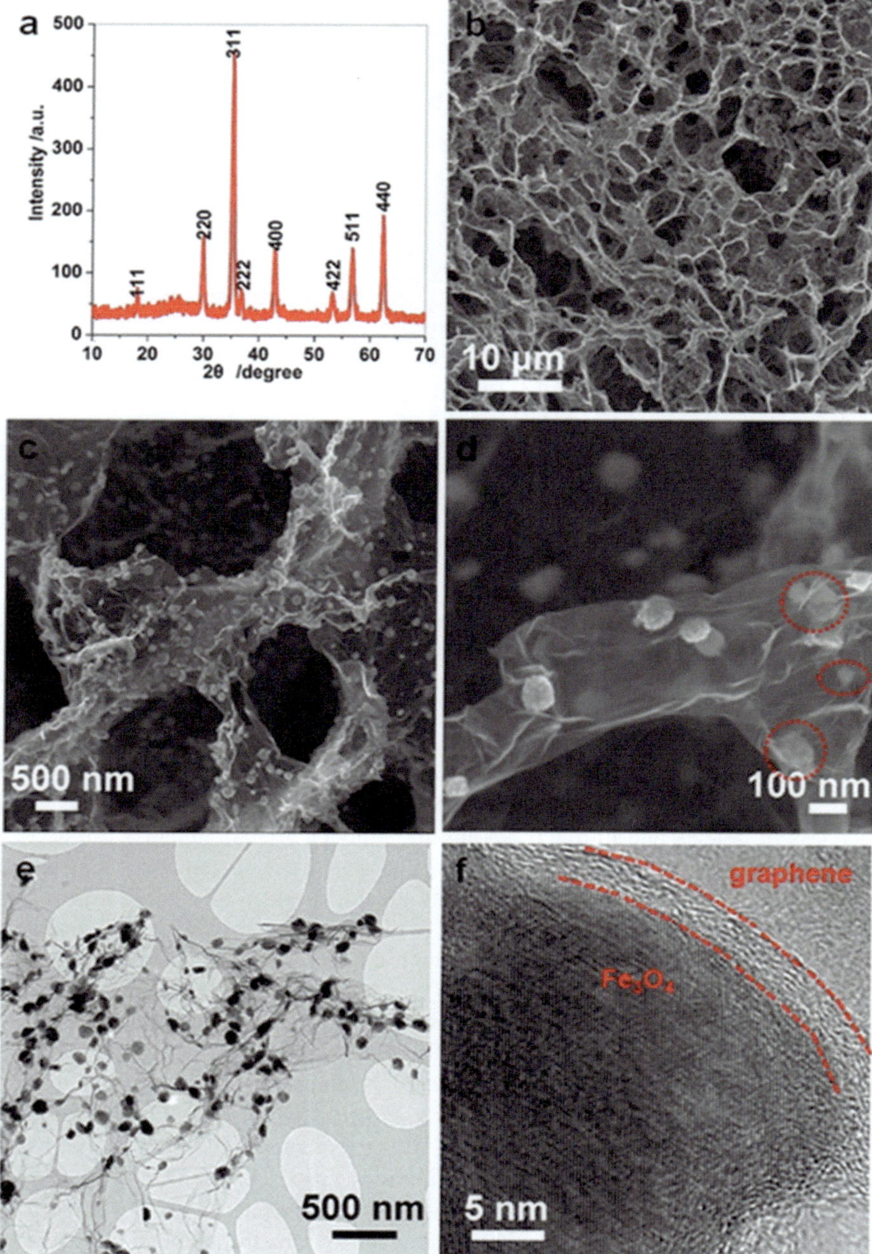

**Fig. 4.12** Structure and morphology of $Fe_3O_4$/N–rGO catalysts. (**a**) XRD pattern and (**b–d**) typical SEM images of $Fe_3O_4$/N–rGOs revealing the 3D macroporous structure and uniform distribution of $Fe_3O_4$ NPs in the GAs. The red rings in (**d**) indicate $Fe_3O_4$ NPs encapsulated in thin graphene layers. Representative (**e**) TEM and (**f**) HRTEM images of $Fe_3O_4$/N–rGO revealing an $Fe_3O_4$ nanoparticles rapped by graphene layers. Reprinted with permission from Ref. [47]. Copyright (2012) American Chemical Society

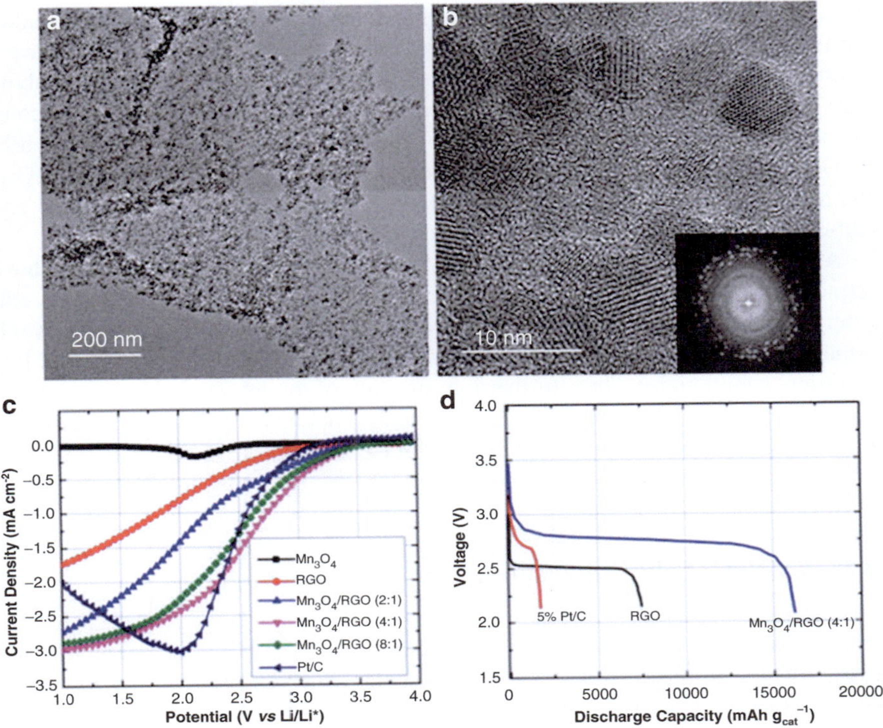

**Fig. 4.13** (**a, b**) SEM and TEM images of the as-prepared $Mn_3O_4$/rGO nanocomposites with an rGO to $Mn_3O_4$ mass ratio of 4:1. (**c**) ORR steady-state RDE polarization curves for various catalysts in $O_2$-saturated 0.1 M $LiPF_6$ in the DME electrolyte. Rotating speed: 900 rpm; room temperature; (**d**) initial discharge performance of various catalysts at a current density of 50 $mAg^{-1}$ in Li–$O_2$ battery tests. Reprinted from Ref. [56] by permission of the Royal Society of Chemistry

media, when compared to $Fe_3O_4$ nanoparticles supported on N-doped carbon black or N-doped rGO sheets. This attests the importance of the 3D macropores and high specific surface area of the GA support for improving the ORR performance. Importantly, better durability was observed with the $Fe_3O_4$/N–rGO, relative to the commercial Pt/C catalyst. This synthetic effect can be further extended to develop other 3D metal or metal oxide/rGO-based monolithic materials for various applications, such as sensors, batteries, and supercapacitors [47].

Apart from the potential application in aqueous for fuel cell application, such rGO/oxide hybrid catalysts also have attracted increasing interests for the applications in nonaqueous Li–$O_2$ batteries [49–58]. Recently, we developed a method for one-step synthesis of $Mn_3O_4$/rGO nanocomposites for nonaqueous Li–$O_2$ batteries [56]. After 24 h of solvothermal reaction, monodispersed $Mn_3O_4$ nanoparticles were formed and uniformly bonded on the surface of rGO (Fig. 4.13a). The $Mn_3O_4$ nanoparticles with diameters of ca. Four to six nanometers are well crystallized (Fig. 4.13b).

Because the monodispersed $Mn_3O_4$ nanoparticles in our composite have smaller particle size compared to previously reported $MnO_x$-based composite, it may have a stronger interaction bonding between $MnO_x$ and rGO and lead to higher catalytic activity accordingly [59]. To investigate the cathode catalyst performance, the composite was evaluated by steady-state RDE and Li–$O_2$ battery test with 1.0 M $LiPF_6$ TEGDME electrolyte (Fig. 4.13c, d). Compared to other ratios of $Mn_3O_4$ and rGO, the $Mn_3O_4$/rGO (4:1) catalyst exhibited highest ORR activity in both half-cell and full-cell test, exhibiting an initial discharge capacity as high as 16 000 mAh g$^{-1}$.

As a new class of ORR catalyst, rGO/transition metal-compound hybrids have attracted more and more attention. However, to date, most of these catalysts still offer inferior performance to that of the Fe–N–rGO catalysts as discussed above. It is partially due to their insufficient inherent activity of oxides or sulfide and low electrical conductivity. Thus, further optimization of the interaction between metal compounds and rGO can further enhance catalyst activity. Noteworthy, these rGO/metal oxide/sulfide catalysts hold great promise to be efficient ORR/OER bifunctional cathode, due to their well-known high OER activity and stability.

## 4.2 GO/rGOs in Supercapacitors

In addition to their versatile applications in catalysis, GO/rGOs also demonstrated extraordinary promises for energy storage devices such as EDLCs (also known as supercapacitors or ultracapacitors) in the last decade [60–65]. Supercapacitors are relatively new energy storage devices as compared to batteries, conventional electrostatic capacitors, and electrolytic capacitors, adopting unique working principle different from any of the above. As shown in Fig. 4.14, in typical electrostatic

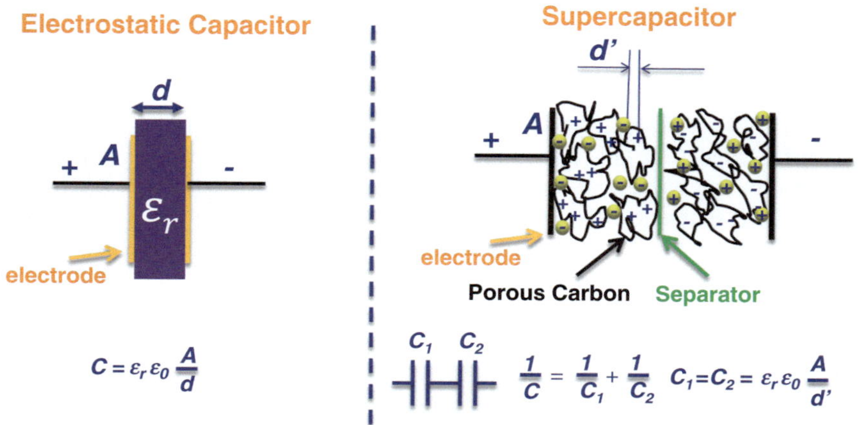


**Fig. 4.14** Comparison of different working mechanisms between supercapacitors and conventional electrostatic capacitors

capacitors, two metal plates are separated by a dielectric. The resulted capacitance can be defined as

$$C = \varepsilon_r \varepsilon_0 \frac{A}{d} \tag{4.1}$$

where $C$ represents capacitance, $\varepsilon r$ represents relative static permittivity, $\varepsilon_0$ represents electric constant ($8.854 \times 10^{-12}$ F/m), $A$ represents the overlapping area of the two plates, and $d$ represents the separation between two plates. However, in the case of supercapacitors, porous carbon materials, such as activated carbons, are used as electrodes, and a macroporous polymer membrane, such as cellulose membranes, is used as the separator. Electrolytes are subsequently added into supercapacitors, in order to form "electric double layers" on porous carbon surfaces. Orders of magnitude higher capacitance is thus generated within the "double layers," which can again be defined by Eq. (4.1). Herein, $A$ becomes the active surface area of porous carbon electrodes ranging from 500 to 3,000 m$^2$/g. $d$ refers to the separation within the "electric double layers" reported to be between 0.5 and 0.8 nm [66]. A real supercapacitor thus can be viewed as a series connection of two "electric double-layer" capacitors. Thanks to the huge $A$ and minimum $d$ in this structure, commercialized supercapacitors usually have capacitance much higher than their electrostatic counter parts. For example, *Maxwell Technologies* have ultracapacitor cells with capacitance as high as 3,400 F ready for sale. Therefore, EDLCs are important complimentary to batteries in existing energy storage technologies, with superior power density, excellent cycling (>10,000 cycles), but inferior energy density as compared to lithium-ion batteries.

We note that, besides the "electric double-layer capacitance" mentioned here, some supercapacitors also adopt another capacitance mechanism, so-called pseudo-capacitance or redox capacitance, which originates from redox reactions of electro-chemically active species in electrodes. For example, redox reactions of transition metal cations, such as Mn, Co, or Ru, as well as some surface oxidation of carbons, have been well investigated in this category [67]. Since redox capacitances actually involve faradic processes, they are similar to the electrochemical processes in batteries.

Since 2008, *Ruoff*'s group at University of Texas, Austin, first introduced rGO as active electrodes in supercapacitors [60]. Hydrazine-reduced GO possesses large surface area (theoretical value of 2,630 m$^2$/g) and good electrical conductivity, which are two key features for an effective electrode material in supercapacitors. The authors reported specific capacitances of 135 and 99 F/g in aqueous and organic electrolytes, respectively (Fig. 4.15) [60]. Three years later, the same research group optimized their recipe of modifications of rGO and obtained a carbon material with surface area as high as 3,100 m$^2$/g (higher than the theoretical value) and electrical conductivity around 500 S/m. The as-made supercapacitor with the high-surface-area rGO achieved an energy density comparable to lead-acid batteries and a power density an order of magnitude higher than that of the commercial carbon-based supercapacitors [61]. We note that the authors were able to obtain an activated material with effective surface area exceeding the theoretical limit of that of graphene. The major explanation for this is that, with KOH etching treatment, large amounts of

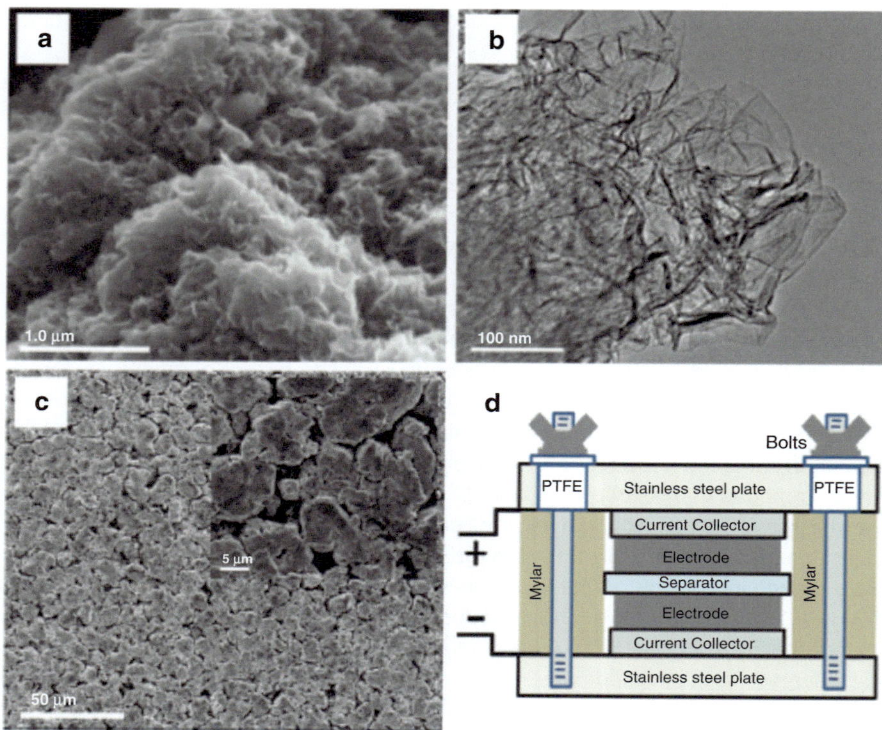


**Fig. 4.15** (**a**) SEM image of CMG particle surface, (**b**) TEM image showing individual graphene sheets extending from CMG particle surface, (**c**) low- and high (*inset*)-magnification SEM images of CMG particle electrode surface, and (**d**) schematic of test cell assembly (Reprinted with permission from Ref [60]. Copyright (2008) American Chemical Society)

pinholes were generated on the graphene basal planes, thus significantly increasing the number of active sites for electric double-layer capacitance in the devices.

Later, laser-induced reduction of GO has been heavily exploited by few research groups, leading to fabrication of rGO-based EDLCs with desired geometries and sizes [62–64]. $CO_2$ laser-patterning technique is widely used in materials engineering for cutting large pieces of plastics or metals. In 2011, Gao et al. first demonstrated the patterning of a freestanding GO film with a $CO_2$ laser cutter (X660, Universal Laser Systems). The as-made patterns on GO worked instantly as supercapacitors in humid environment (Fig. 4.16) [62]. It is important to note that in this work, the authors also discovered the proton-conducting behavior of hydrated GO films, which was later explored as a proton-exchange membrane in polymer electrolyte fuel cells [68].

Soon after, Kaner's group in University of California at Los Angeles demonstrated similar fabrication processes of EDLCs with a LightScribe DVD optical drive as the reduction tool for GO in 2012 [63, 64]. They coated a thin layer of GO onto a DVD disk and light-scribed it in a computerized DVD drive. The resulted film was peeled

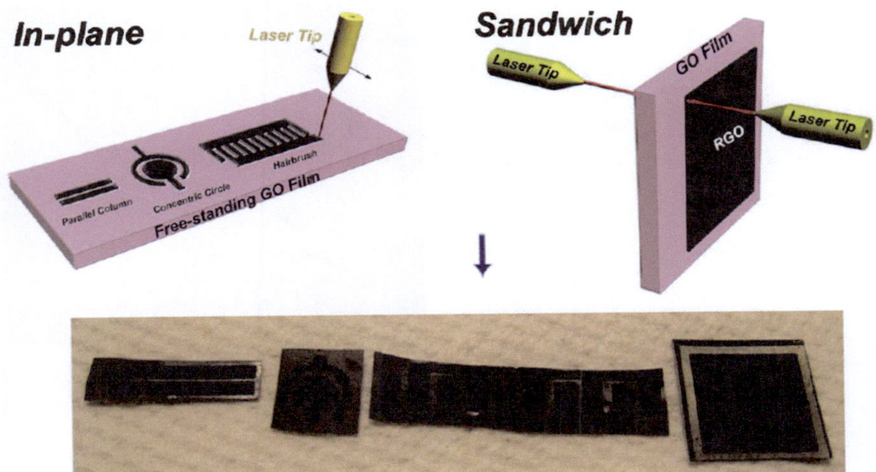


**Fig. 4.16** Schematics of $CO_2$ laser patterning of freestanding hydrated GO films to fabricate RGO–GO–RGO devices with in-plane and sandwich geometries. The *black contrast* in the *top schematics* corresponds to RGO and the light contrast to unmodified hydrated GO. For in-plane devices, three different geometries were used, and the concentric circular pattern gives the highest capacitance density. The *bottom row* shows photographs of patterned films (Reprinted with permission from Macmillan Publishers Ltd: Nature Nanotechnology [62], copyright 2011)

off and assembled into a sandwich supercapacitor structure. They reported a surface area of 1,520 $m^2$/g and an electrical conductivity of 1,738 S/m [63]. The as-made devices can deliver a power density of ~20 $W/cm^3$, 20 times higher than that of activated-carbon-based EDLCs [63]. Later, the same group optimized their laser treating recipe and demonstrated hundreds of smaller supercapacitors produced on a single DVD disk within 30 min and power densities of ~200 $W/cm^3$, among the highest value achieved for supercapacitors then (Fig. 4.17) [64]. There are two notable aspects of their work worthy of a closer look. The first one is that the supercapacitors they demonstrated don't require any current collectors, as opposite to most of other reports on supercapacitors. This is mainly due to the high electrical conductivity measured for their laser-treated rGO samples. The second point is about the flexibility of their as-fabricated devices. These devices can be bent and twisted several times without any performance loss. These fascinating properties are definitely important breakthrough in supercapacitor research, and their products offered great feasibility to integrate with MEMS or CMOS in a single chip [64].

In 2013, Dan Li and coworkers in Monash University, Australia, made further improvements on rGO-based ultracapacitors [65, 69]. They successfully prepared rGO films with high ion-accessible surface area and low ion transport resistance, with the help of a capillary compression process in rGO-gel films in the presence of a nonvolatile liquid electrolyte. To be more specific, they followed their previous recipe to make rGO hydrogel films [69] with hydrazine reduction (Fig. 4.18). Then the films were soaked in a ratio-controlled volatile/nonvolatile miscible solutions,

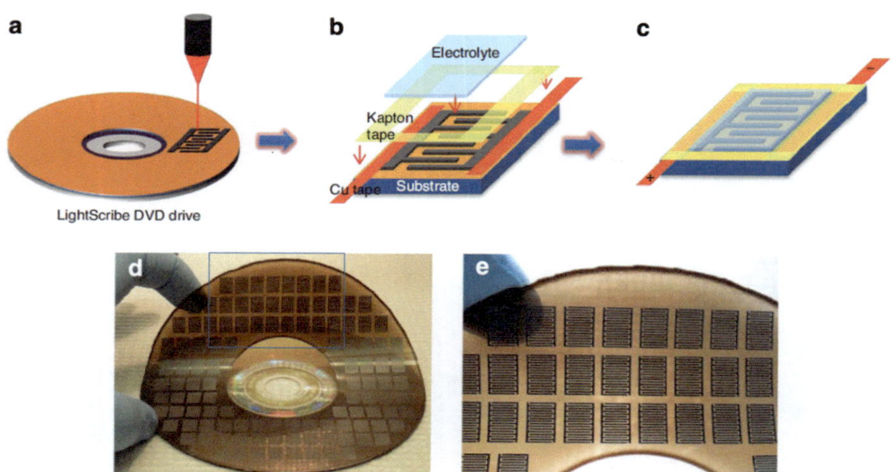



**Fig. 4.17** Fabrication of LSG-MSL. (**a–c**) Schematic diagram showing the fabrication process for an LSG micro-supercapacitor. A GO film supported on a PET sheet is placed on a DVD media disk. The disk is inserted into a LightScribe DVD drive and a computer-designed circuit is etched onto the film. (**a**) The laser inside the drive converts the *golden-brown* GO into black LSG at precise locations to produce interdigitated graphene circuits. (**b**) Copper tape is applied along the edges to improve the electrical contacts, and the interdigitated area is defined by polyimide (Kapton) tape. (**c**) An electrolyte overcoat is then added to create a planar micro-supercapacitor. (**d, e**) This technique has the potential for the direct writing of micro-devices with high areal density. More than 100 micro-devices can be produced on a single run. The micro-devices are completely flexible and can be produced on virtually any substrate (Reprinted with permission from Macmillan Publishers Ltd: Nature Communications [64], copyright (2013))

stirred for 12 h to reach equilibrium. Deionized water was used as the volatile liquids, whereas sulfuric acid and an ionic liquid were used as two nonvolatile liquids. When they had a solvent-exchanged rGO hydrogel film, they placed it in a vacuum oven and the volatile liquid was removed under high vacuum (10 Pa). The evaporation of water exerted capillary compression between rGO layers and led to shrinkage of the thickness and increase of packing density. Superior volumetric energy density (59.9 Wh/L) and power density (8.6 kW/L) were achieved in their devices, mainly due to the much higher packing density of their electrode–electrolyte system. The key feature of their system is that the preincorporation of electrolytes in the rGO films ensured excellent rate performance of their supercapacitors.

In summary, rGO-based EDLCs are of great importance to both scientific and technological communities. Major breakthroughs have been demonstrated in the past 5 years; however, the biggest issues in EDLC research remain to be unsolved. For example, the energy density of EDLCs is still lower than the widely used lithium-ion batteries in the market. Furthermore, few researchers have investigated the self-discharge processes in supercapacitors, which are known to be orders of magnitude faster than that in batteries. To make ultracapacitors competitive alternatives to batteries, these technological issues will have to be addressed.

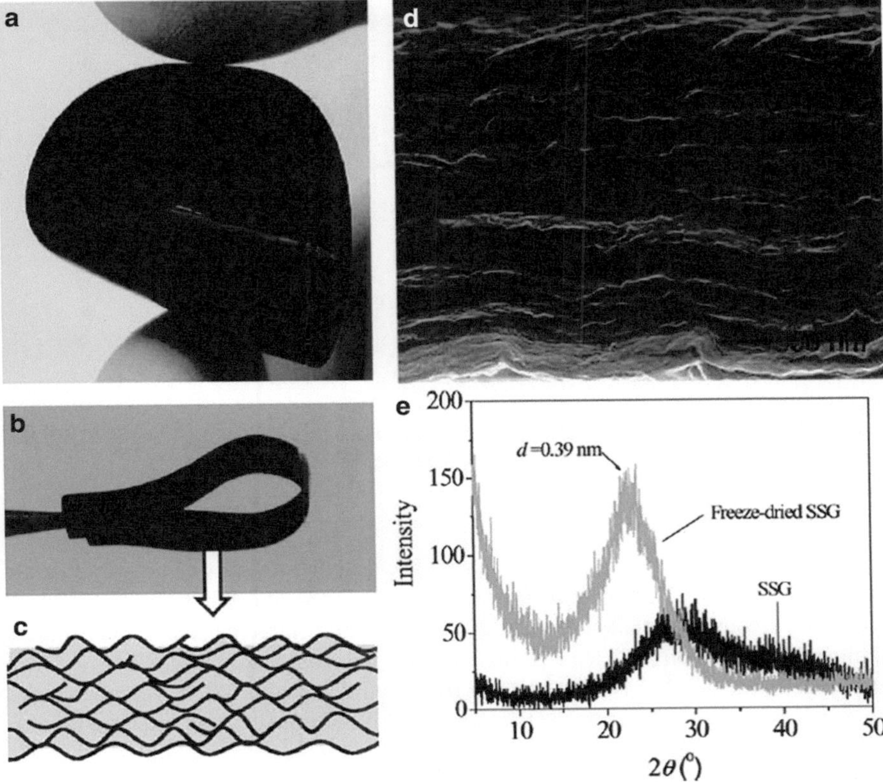



**Fig. 4.18** Structure of the SSG film. (**a**, **b**) Photographs of the as-formed flexible SSG films. (**c**) Schematic of the cross section of the SSG film. (**d**) SEM image of the cross section of a freeze-dried SSG film. Given that the undried SSG film must be much more porous than shown in this SEM image. (**e**) XRD patterns of as-prepared and freeze-dried SSG films. The broad diffraction peak of the wet SSG film (from 27.9 to 44.8°) is likely due to the water confined in the CCG network because the corresponding d-spacing is less than the $d_{(002)}$ of graphite. Further experiments are required to understand the structure of the water in this SSG film (Reprinted with permission from Ref [69], copyright (2011) John Wiley & Sons, Inc.)

## 4.3 GO/rGO Anodes in Li-Ion Batteries

Today, Li-ion batteries (LIBs) have become the most promising energy storage technologies for portable electronics and transportation. However, the energy density of current batteries is still not enough to provide required energy after a single charge. The challenges are related to insufficient capacity and power density of electrodes as well as high cost and serious safety issue. In order to overcome these limitations, advanced electrode materials are required. Current LIBs use layered graphite materials as anodes, capable of storing ions in the material's bulk, but

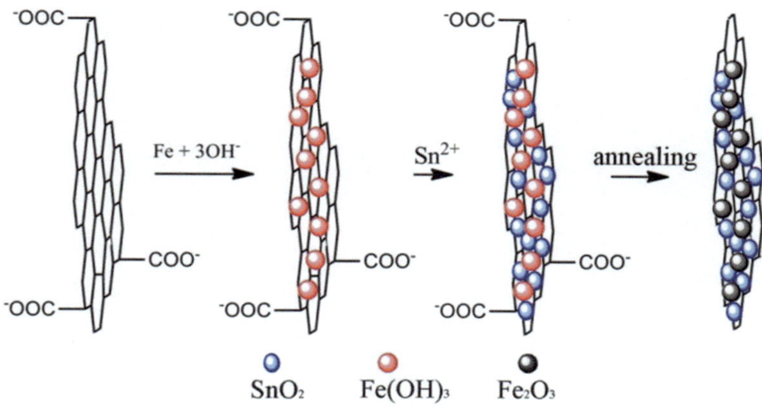



**Fig. 4.19** Scheme of preparation of ternary rGO/Fe$_2$O$_3$/SnO$_2$nanocompoiste. Reprinted with permission from Ref. [93]. Copyright (2013) American Chemical Society

suffer from low energy and power density. The graphite anode can only intercalate one Li atom per six carbon atoms (LiC$_6$) with a theoretical capacity of 372 mAhg$^{-1}$ [70]. New anode materials with fast electron transport, large capacity, and efficient lithium-ion diffusion will lead to high-power and high-rate LIBs [71–74]. Thus, attention on the anode has been shifting from traditional graphite to other advanced materials with an aim to increase the number of lithiation sites and improve the diffusivity of Li [75–78].

Graphene materials, possessing high electron conductivity, large specific surface area (up to 2,600 m$^2$/g), and a broad window of electrochemical stability, hold great promise as an advanced material for energy storage technologies [21, 79–81]. The specific capacity of graphene for Li can be substantially higher than that of graphite, because graphene can adsorb lithium on both sides. Furthermore, the single layer of graphene provides a facile route for the diffusion of lithium ions, since the space for lithium intercalation is much larger than that in graphite interlayers [81–84]. Despite these advantages, graphene anodes experience significant irreversible capacity losses during charge/discharge cycling, mainly due to the restacking of graphene layers [85]. Recently, it was found that this problem can be alleviated by incorporating solid nanoparticles (e.g., Si, CuO, Fe$_2$O$_3$, SnO$_2$ Co$_3$O$_4$, or Mn$_3$O$_4$) in between the sheets to reduce the restacking degree [86–92]. Here, only one example is given based on our own research.

Recently, we proposed a novel approach to designing and synthesizing a ternary rGO/Fe$_2$O$_3$/SnO$_2$ graphene nanocomposite consisting of rGO with incorporation of highly Li active Fe$_2$O$_3$ and SnO$_2$ particles [93]. As shown in Fig. 4.19 [93], the rGO/Fe$_2$O$_3$/SnO$_2$ ternary nanocomposite was synthesized via an in situ precipitation of Fe$_2$O$_3$ nanoparticles using FeCl$_3$ precursor onto GO, followed by a subsequent reduction with SnCl$_2$ to obtain highly conductive rGO.

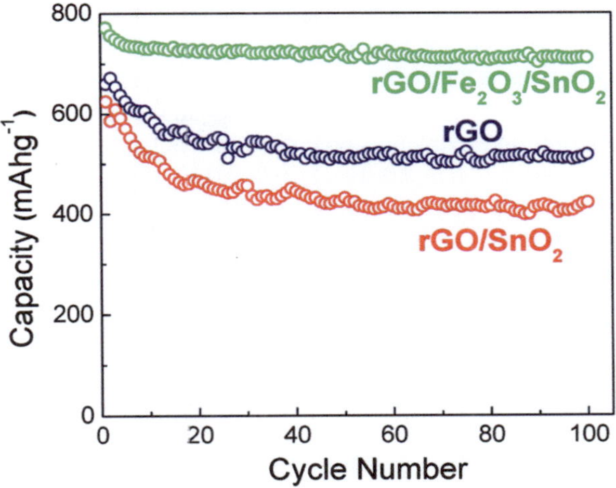


**Fig. 4.20** A comparison of discharge–charge capacity measured with rGO, rGO/SnO$_2$, and rGO/Fe$_2$O$_3$/SnO$_2$ anodes up to 100 cycles. Reprinted with permission from Ref. [93]. Copyright (2013) American Chemical Society

Typically, GO aqueous solution was prepared through a modified Hummers method by using natural graphite powder as a precursor [94, 95]. In a typical synthesis procedure, FeCl$_3$·6H$_2$O (0.8 g) and urea (1.5 g) were added to 150 mL of GO solution (containing about 0.12 g GO) under constant stirring. After sonicating for 30 min, the suspension was transferred to a Teflon-lined autoclave and maintained at 120 °C for 4 h. Because the oxygen functional groups on the GO sheets can chemically bond with the metals ions, Fe$^{3+}$ ions were likely anchored on the planes of the GO sheets [96]. Therefore, Fe$^{3+}$ can be precipitated by OH$^-$ on GO sheets when the urea decomposes during the hydrothermal process. In order to realize the reduction of GO, the suspension was adjusted to pH=7.0 after cooling to room temperature. Then, SnCl$_2$·H$_2$O was introduced into the above solution and a temperature of 30 °C was maintained for 1.0 h with continuous magnetic stirring to achieve a maximum reduction efficiency. The resulting precipitation was washed with water and ethanol several times during the centrifuge separation process.

As shown in Fig. 4.20 [93], the ternary anode material is superior to rGO and binary rGO/SnO$_2$ anodes, showing improved cycle stability and specific capacity. After 100 cycles, the charge capacities measured with the rGO/Fe$_2$O$_3$/SnO$_2$ anode are the highest relative to those of rGO and rGO/SnO$_2$, retaining a specific capacity of above 700 mAhg$^{-1}$ which can be retained during the subsequent 100 cycles. In the rGO-based ternary anode materials, rGO, Fe$_2$O$_3$, and SnO$_2$ particles play different but complementary roles. rGO serves as a matrix enabling both lithium ions and electrons to migrate to active sites, thereby fully maximizing the energy density offered by both graphene and Fe$_2$O$_3$. The rGO also is an effective elastic buffer to relieve the strain that would otherwise accumulate in the agglomerated Fe$_2$O$_3$ particles during Li uptake/release.

In turn, the dispersion of $Fe_2O_3$ and $SnO_2$ particles on graphene prevents the restacking of graphene sheets, maintaining a high storage capacity of lithium during cycling. It is worth noting that the rGO used in this work is prepared via reduction of GO by $SnCl_2$, a nontoxic reducing agent, introducing $SnO_2$ in the same step. However, both the lithium insertion and extraction potentials of $SnO_2$ are below 0.7 V and just beyond the active potential range of $Fe_2O_3$ (0.7–2.0 V). Thus, due to the discrepancy in electrochemical active potentials, $SnO_2$ is expected to mainly serve as an inert matrix for $Fe_2O_3$ particles and keep them from agglomerating. This might be an important reason for the high performance observed from the graphene nanocomposite anode during the discharge–charge cycling.

Future research of rGO-based anodes will focus on (1) gaining a fundamental understanding of lithium-ion adsorption/desorption and diffusion kinetics in heteroatom-doped graphene structures using the proposed model systems and density functional theory (DFT) calculations, (2) exploring the facilitating roles of graphene and Si particles and their possible synergistic effects in developed composite anode materials, and (3) designing and synthesizing graphene-based composites with controllable morphology and structure for advanced lithium battery anodes. The cost-effective and functional rGO-based anodes will substantially improve $Li^+$-specific capacity, mass transport, and charge/discharge cyclic stability for next-generation LIBs.

## 4.4   Summary and Perspective

In the past decade, GOs or rGOs have emerged as important energy materials for electrochemical energy storages and conversion technologies, including fuel cells, metal–air batteries, supercapacitors, and lithium-ion batteries. Due to unique properties of rGO including high surface area, excellent electrical conductivity, good electrochemical stability, and easy functionalization of two-dimensional planes, superior performance was universally achieved on these graphene-based catalysts or electrode materials for these electrochemical devices. In this chapter, applications of GO/rGO-based materials were selectively reviewed in terms of their use as nonprecious metal catalysts in fuel cells and metal–air batteries and as high-performance electrodes in supercapacitors and lithium-ion batteries. The purpose is to highlight the comparisons among material synthesis, processing, structures, and properties.

Generally, during the synthesis of rGO, graphite precursors, reducing agents, and methods are critical to resulting morphologies and structures of rGO. Furthermore, the surface areas, porosity, defects, dopants, and reduction degree of rGO can directly link to corresponding material performance. Among studied methods to modify rGO, heteroatom doping is very effective to tune the structural and electronic properties of carbon planes in rGO, creating more active species for associated electrochemical reactions in various applications. In addition, development of rGO nanocomposites by integrating with nanoparticles and functional polymers will result in possible synergistic effects and further resolve the current problems of rGO materials, thereby significantly enhancing performance metrics. Importantly, understanding the nature of active site or adsorption sites in these rGO materials and the associated reaction mechanism will further provide insights to rational design and synthesis of advanced rGO materials.

**Acknowledgements**  G. W. acknowledges the support from the start-up funding of University at Buffalo, The State University of New York.

W. G. sincerely thank for the start-up funding support from the Department of Textile Engineering, Chemistry and Science at North Carolina State University, Raleigh, NC.

# References


1. Abraham KM, Jiang Z (1996) A polymer electrolyte-based rechargeable lithium/oxygen battery. J Electrochem Soc 143:1–5
2. Van Mierlo J, Maggetto G (2007) Fuel cell or battery: electric cars are the future. Fuel Cells 7:165–173
3. Girishkumar G, McCloskey B, Luntz AC, Swanson S, Wilcke W (2010) Lithium–air battery: promise and challenges. J Phys Chem Lett 1:2193–2203
4. Debart A, Paterson AJ, Bao J, Bruce PG (2008) Alpha-MnO$_2$ nanowires: a catalyst for the O$_2$ electrode in rechargeable lithium batteries. Angew Chem Int Ed 47:4521–4524
5. Wu G, Zelenay P (2013) Nanostructured nonprecious metal catalysts for oxygen reduction reaction. Acc Chem Res 46:1878–1889
6. Li Q, Cao R, Chol J, Wu G (2014) Nanocarbon electrocatalysts for oxygen reduction in alkaline media for advanced energy conversion and storage. Adv Energy Mater 4:1301415
7. Li Q, Cao R, Cho J, Wu G (2014) Nanostructured carbon-based cathode catalysts for nonaqueous lithium-oxygen batteries. Phys Chem Chem Phys 16:13568–13582
8. Calle-Vallejo F, Martinez JI, Rossmeisl J (2011) Density functional studies of functionalized graphitic materials with late transition metals for oxygen reduction reactions. Phys Chem Chem Phys 13:15639–15643
9. Parvez K, Yang SB, Hernandez Y, Winter A, Turchanin A, Feng XL, Mullen K (2012) Nitrogen-doped graphene and its iron-based composite as efficient electrocatalysts for oxygen reduction reaction. ACS Nano 6:9541–9550
10. Wang H, Yang Y, Liang Y, Zheng G, Li Y, Cui Y, Dai H (2012) Rechargeable Li-O$_2$ batteries with a covalently coupled MnCo$_2$O$_4$-graphene hybrid as an oxygen cathode catalyst. Energy Environ Sci 5:7931–7935
11. Wu G, Nelson MA, Mack NH, Ma SG, Sekhar P, Garzon FH, Zelenay P (2010) Titanium dioxide-supported non-precious metal oxygen reduction electrocatalyst. Chem Commun 46:7489–7491
12. Wu G, More KL, Xu P, Wang H-L, Ferrandon M, Kropf AJ, Myers DJ, Ma S, Johnston CM, Zelenay P (2013) A carbon-nanotube-supported graphene-rich non-precious metal oxygen reduction catalyst with enhanced performance durability. Chem Commun 49:3291–3293
13. Zhang S, Dai LM (2013) Metal-free electrocatalysts for oxygen reduction. In: Shao M (ed) Electrocatalysis in fuel cells. Springer, London, pp 375–390
14. Gong KP, Du F, Xia ZH, Durstock M, Dai LM (2009) Nitrogen-doped carbon nanotube arrays with high electrocatalytic activity for oxygen reduction. Science 323:760–764
15. Qu LT, Liu Y, Baek JB, Dai LM (2010) Nitrogen-doped graphene as efficient metal-free electrocatalyst for oxygen reduction in fuel cells. ACS Nano 4:1321–1326
16. Yeager E (1984) Electrocatalysts for O$_2$ reduction. Electrochim Acta 29:1527–1537
17. Wu G, More KL, Johnston CM, Zelenay P (2011) High-performance electrocatalysts for oxygen reduction derived from polyaniline, iron, and cobalt. Science 332:443–447

18. Wu G, Li DY, Dai CS, Wang DL, Li N (2008) Well-dispersed high-loading Pt nanoparticles supported by shell-core nanostructured carbon for methanol electrooxidation. Langmuir 24:3566–3575
19. Pels JR, Kapteijn F, Moulijn JA, Zhu Q, Thomas KM (1995) Evolution of nitrogen functionalities in carbonaceous materials during pyrolysis. Carbon 33:1641–1653
20. Wu G, Dai C, Wang D, Li D, Li N (2010) Nitrogen-doped magnetic onion-like carbon as support for Pt particles in a hybrid cathode catalyst for fuel cells. J Mater Chem 20:3059–3068
21. Wu G, Mack NH, Gao W, Ma S, Zhong R, Han J, Baldwin JK, Zelenay P (2012) Nitrogen-doped graphene-rich catalysts derived from heteroatom polymers for oxygen reduction in non-aqueous lithium–$O_2$ battery cathodes. ACS Nano 6:9764–9776
22. Wu G, Swaidan R, Li D, Li N (2008) Enhanced methanol electro-oxidation activity of PtRu catalysts supported on heteroatom-doped carbon. Electrochim Acta 53:7622–7629
23. Sidik RA, Anderson AB, Subramanian NP, Kumaraguru SP, Popov BN (2006) $O_2$ reduction on graphite and nitrogen-doped graphite: experiment and theory. J Phys Chem B 110:1787–1793
24. Kim H, Lee K, Woo SI, Jung Y (2011) On the mechanism of enhanced oxygen reduction reaction in nitrogen-doped graphene nanoribbons. Phys Chem Chem Phys 13:17505–17510
25. Li XL, Wang HL, Robinson JT, Sanchez H, Diankov G, Dai HJ (2009) Simultaneous nitrogen doping and reduction of graphene oxide. J Am Chem Soc 131:15939–15944
26. Geng DS, Chen Y, Chen YG, Li YL, Li RY, Sun XL, Ye SY, Knights S (2011) High oxygen-reduction activity and durability of nitrogen-doped graphene. Energy Environ Sci 4:760–764
27. Li N, Wang ZY, Zhao KK, Shi ZJ, Gu ZN, Xu SK (2010) Synthesis of single-wall carbon nanohorns by arc-discharge in air and their formation mechanism. Carbon 48:1580–1585
28. Panchokarla LS, Subrahmanyam KS, Saha SK, Govindaraj A, Krishnamurthy HR, Waghmare UV, Rao CNR (2009) Synthesis, structure, and properties of boron- and nitrogen-doped graphene. Adv Mater 21:4726–4730
29. Shao YY, Zhang S, Engelhard MH, Li GS, Shao GC, Wang Y, Liu J, Aksay IA, Lin YH (2010) Nitrogen-doped graphene and its electrochemical applications. J Mater Chem 20:7491–7496
30. Jeong HM, Lee JW, Shin WH, Choi YJ, Shin HJ, Kang JK, Choi JW (2011) Nitrogen-doped graphene for high-performance ultracapacitors and the importance of nitrogen-doped sites at basal planes. Nano Lett 11:2472–2477
31. Liang YY, Li YG, Wang HL, Zhou JG, Wang J, Regier T, Dai HJ (2011) $Co_3O_4$ nanocrystals on graphene as a synergistic catalyst for oxygen reduction reaction. Nat Mater 10:780–786
32. Sheng ZH, Shao L, Chen JJ, Bao WJ, Wang FB, Xia XH (2011) Catalyst-Free synthesis of nitrogen-doped graphene via thermal annealing graphite oxide with melamine and its excellent electrocatalysis. ACS Nano 5:4350–4358
33. Lai L, Potts JR, Zhan D, Wang L, Poh CK, Tang C, Gong H, Shen Z, Lin J, Ruoff RS (2012) Exploration of the active center structure of nitrogen-doped graphene-based catalysts for oxygen reduction reaction. Energy Environ Sci 5:7936–7942
34. Sun B, Wang B, Su DW, Xiao LD, Ahn H, Wang GX (2012) Graphene nanosheets as cathode catalysts for lithium-air batteries with an enhanced electrochemical performance. Carbon 50:727–733
35. Xiao J, Mei D, Li X, Xu W, Wang D, Graff GL, Bennett WD, Nie Z, Saraf LV, Aksay IA, Liu J, Zhang J-G (2011) Hierarchically porous graphene as a lithium–air battery electrode. Nano Lett 11:5071–5078
36. Seredych M, Bandosz TJ (2014) Confined space reduced graphite oxide doped with sulfur as metal-free oxygen reduction catalyst. Carbon 66:227–233
37. Liang J, Jiao Y, Jaroniec M, Qiao SZ (2012) Sulfur and nitrogen dual-doped mesoporous graphene electrocatalyst for oxygen reduction with synergistically enhanced performance. Angew Chem Int Ed 51:11496–11500
38. Yang Z, Yao Z, Li G, Fang G, Nie H, Liu Z, Zhou X, Chen XA, Huang S (2011) Sulfur-doped graphene as an efficient metal-free cathode catalyst for oxygen reduction. ACS Nano 6:205–211
39. Wang SY, Iyyamperumal E, Roy A, Xue YH, Yu DS, Dai LM (2011) Vertically aligned BCN nanotubes as efficient metal-free electrocatalysts for the oxygen reduction reaction: a synergetic effect by co-doping with boron and nitrogen. Angew Chem Int Ed 50:11756–11760

40. Sheng Z-H, Gao H-L, Bao W-J, Wang F-B, Xia X-H (2012) Synthesis of boron doped graphene for oxygen reduction reaction in fuel cells. J Mater Chem 22:390–395

41. Byon HR, Suntivich J, Shao-Horn Y (2011) Graphene-based non-noble-metal catalysts for oxygen reduction reaction in acid. Chem Mater 23:3421–3428

42. Liang YY, Wang HL, Zhou JG, Li YG, Wang J, Regier T, Dai HJ (2012) Covalent hybrid of spinel manganese-cobalt oxide and graphene as advanced oxygen reduction electrocatalysts. J Am Chem Soc 134:3517–3523

43. Wu G, Li N, Zhou D-R, Mitsuo K, Xu B-Q (2004) Anodically electrodeposited Co + Ni mixed oxide electrode: preparation and electrocatalytic activity for oxygen evolution in alkaline media. J Solid State Chem 177:3682–3692

44. Wang HL, Liang YY, Li YG, Dai HJ (2011) Co1-xS-graphene hybrid: a high-performance metal chalcogenide electrocatalyst for oxygen reduction. Angew Chem Int Ed 50: 10969–10972

45. Yan XY, Tong XL, Zhang YF, Han XD, Wang YY, Jin GQ, Qin Y, Guo XY (2012) Cuprous oxide nanoparticles dispersed on reduced graphene oxide as an efficient electrocatalyst for oxygen reduction reaction. Chem Commun 48:1892–1894

46. Li Q, Xu P, Zhang B, Tsai H, Zheng S, Wu G, Wang H-L (2013) Structure-dependent electrocatalytic properties of $Cu_2O$ nanocrystals for oxygen reduction reaction. J Phys Chem C 117:13872

47. Wu ZS, Yang SB, Sun Y, Parvez K, Feng XL, Mullen K (2012) 3D Nitrogen-doped graphene aerogel-supported $Fe_3O_4$ nanoparticles as efficient eletrocatalysts for the oxygen reduction reaction. J Am Chem Soc 134:9082–9085

48. He Q, Li Q, Khene S, Ren X, López-Suárez FE, Lozano-Castello D, Bueno-López A, Wu G (2013) High-loading cobalt oxide coupled with nitrogen-doped graphene for oxygen-reduction in anion-exchange membrane alkaline fuel cells. J Phys Chem C 117:8697–8707

49. Lee J-S, Tai Kim S, Cao R, Choi N-S, Liu M, Lee KT, Cho J (2011) Metal–air batteries with high energy density: Li–Air versus Zn–Air. Adv Energy Mater 1:34–50

50. Cheng F, Chen J (2012) Metal-air batteries: from oxygen reduction electrochemistry to cathode catalysts. Chem Soc Rev 41:2172–2192

51. Trahey L, Johnson CS, Vaughey JT, Kang S-H, Hardwick LJ, Freunberger SA, Bruce PG, Thackeray MM (2011) Activated lithium-metal-oxides as catalytic electrodes for $Li–O_2$ cells. Electrochem Solid State Lett 14:A64–A66

52. Zhao Y, Xu L, Mai L, Han C, An Q, Xu X, Liu X, Zhang Q (2012) Hierarchical mesoporous perovskite La0.5Sr0.5CoO2.91 nanowires with ultrahigh capacity for Li-air batteries. Proc Natl Acad Sci U S A 109:19569–19574

53. Oh SH, Nazar LF (2012) Oxide catalysts for rechargeable high-capacity $Li–O_2$ batteries. Adv Energy Mater 2:903–910

54. Grimaud A, Carlton CE, Risch M, Hong WT, May KJ, Shao-Horn Y (2013) Oxygen evolution activity and stability of $Ba_6Mn_5O_{16}$, $Sr_4Mn_2CoO_9$, and Sr6Co5O15: the influence of transition metal coordination. J Phys Chem C 117:25926–25932

55. Jung KN, Lee JI, Im WB, Yoon S, Shin KH, Lee JW (2012) Promoting $Li_2O_2$ oxidation by an La(1.7)Ca(0.3)Ni(0.75)Cu(0.25)O4 layered perovskite in lithium-oxygen batteries. Chem Commun (Camb) 48(75):9406–9408

56. Shao YY, Ding F, Xiao J, Zhang J, Xu W, Park S, Zhang JG, Wang Y, Liu J (2013) Making Li-air batteries rechargeable: material challenges. Adv Funct Mater 23:987–1004

57. Cao R, Lee JS, Liu ML, Cho J (2012) Recent Progress in non-precious catalysts for metal-air batteries. Adv Energy Mater 2:816–829

58. Wang HL, Yang Y, Liang YY, Zheng GY, Li YG, Cui Y, Dai HJ (2012) Rechargeable $Li-O_2$ batteries with a covalently coupled $MnCo_2O_4$-graphene hybrid as an oxygen cathode catalyst. Energy Environ Sci 5:7931–7935

59. Wang H, Dai H (2013) Strongly coupled inorganic-nano-carbon hybrid materials for energy storage. Chem Soc Rev 42:3088–3113

60. Stoller MD, Park S, Zhu Y, An J, Ruoff RS (2008) Graphene-based ultracapacitors. Nano Lett 8:3498–3502

61. Zhu Y, Murali S, Stoller MD, Ganesh K, Cai W, Ferreira PJ, Pirkle A, Wallace RM, Cychosz KA, Thommes M (2011) Carbon-based supercapacitors produced by activation of graphene. Science 332:1537–1541

62. Gao W, Singh N, Song L, Liu Z, Reddy ALM, Ci L, Vajtai R, Zhang Q, Wei B, Ajayan PM (2011) Direct laser writing of micro-supercapacitors on hydrated graphite oxide films. Nat Nanotechnol 6:496–500

63. El-Kady MF, Strong V, Dubin S, Kaner RB (2012) Laser scribing of high-performance and flexible graphene-based electrochemical capacitors. Science 335:1326–1330

64. El-Kady MF, Kaner RB (2013) Scalable fabrication of high-power graphene micro-supercapacitors for flexible and on-chip energy storage. Nat Commun 4:1475

65. Yang X, Cheng C, Wang Y, Qiu L, Li D (2013) Liquid-mediated dense integration of graphene materials for compact capacitive energy storage. Science 341:534–537

66. Namisnyk AM (2003) A survey of electrochemical supercapacitor technology. University of Technology, Sydney

67. Reddy ALM, Ramaprabhu S (2007) Nanocrystalline metal oxides dispersed multiwalled carbon nanotubes as supercapacitor electrodes. J Phys Chem C 111:7727–7734

68. Gao W, Wu G, Janicke MT, Cullen DA, Mukundan R, Baldwin JK, Brosha EL, Galande C, Ajayan PM, More KL, Dattelbaum AM, Zelenay P (2014) Ozonated Graphene Oxide Film as a Proton-Exchange Membrane. Angew Chem Int Ed 53:3588–3593

69. Yang X, Zhu J, Qiu L, Li D (2011) Bioinspired effective prevention of restacking in multilayered graphene films: towards the next generation of high-performance supercapacitors. Adv Mater 23:2833–2838

70. Dahn JR, Zheng T, Liu Y, Xue JS (1995) Mechanisms for lithium insertion in carbonaceous materials. Science 270:590–593

71. Chan CK, Peng H, Liu G, McIlwrath K, Zhang XF, Huggins RA, Cui Y (2008) High-performance lithium battery anodes using silicon nanowires. Nat Nanotechnol 3:31–35

72. Armand M, Tarascon JM (2008) Building better batteries. Nature 451:652–657

73. Chen Z, Dai C, Wu G, Nelson M, Hu X, Zhang R, Liu J, Xia J (2010) High performance $Li_3V_2(PO_4)_3$/C composite cathode material for lithium ion batteries studied in pilot scale test. Electrochim Acta 55:8595–8599

74. Xu X, Cao R, Jeong S, Cho J (2012) Spindle-like mesoporous $\alpha$-$Fe_2O_3$ anode material prepared from MOF template for high-rate lithium batteries. Nano Lett 12:4988–4991

75. Han F-D, Bai Y-J, Liu R, Yao B, Qi Y-X, Lun N, Zhang J-X (2011) Template-free synthesis of interconnected hollow carbon nanospheres for high-performance anode material in lithium-ion batteries. Adv Energy Mater 1:798–801

76. Chan CK, Patel RN, O'Connell MJ, Korgel BA, Cui Y (2010) Solution-grown silicon nanowires for lithium-ion battery anodes. ACS Nano 4:1443–1450

77. Zhu X, Zhu Y, Murali S, Stoller MD, Ruoff RS (2011) Nanostructured reduced graphene oxide/$Fe_2O_3$ composite as a high-performance anode material for lithium ion batteries. ACS Nano 5:3333–3338

78. Liu R, Li N, Xia G, Li D, Wang C, Xiao N, Tian D, Wu G (2013) Assembled hollow and core-shell $SnO_2$ microspheres as anode materials for Li-ion batteries. Mater Lett 93:243–246

79. Stankovich S, Dikin DA, Dommett GHB, Kohlhaas KM, Zimney EJ, Stach EA, Piner RD, Nguyen ST, Ruoff RS (2006) Graphene-based composite materials. Nature 442:282–286

80. Yoo E, Kim J, Hosono E, Zhou H-S, Kudo T, Honma I (2008) Large reversible Li storage of graphene nanosheet families for use in rechargeable lithium ion batteries. Nano Lett 8:2277–2282

81. Wang C, Li D, Too CO, Wallace GG (2009) Electrochemical properties of graphene paper electrodes used in lithium batteries. Chem Mater 21:2604–2606

82. Lian P, Zhu X, Liang S, Li Z, Yang W, Wang H (2010) Large reversible capacity of high quality graphene sheets as an anode material for lithium-ion batteries. Electrochim Acta 55:3909–3914

83. Wang G, Shen X, Yao J, Park J (2009) Graphene nanosheets for enhanced lithium storage in lithium ion batteries. Carbon 2049–2053:47

84. Wu Z-S, Ren W, Xu L, Li F, Cheng H-M (2011) Doped Graphene Sheets As Anode Materials with Superhigh Rate and Large Capacity for Lithium Ion Batteries. ACS Nano 5:5463–5471
85. Evanoff K, Magasinski A, Yang J, Yushin G (2011) Nanosilicon-Coated Graphene Granules as Anodes for Li-Ion Batteries. Adv Energy Mater 1:495–498
86. Wang H, Cui L-F, Yang Y, Sanchez Casalongue H, Robinson JT, Liang Y, Cui Y, Dai H (2010) $Mn_3O_4$-graphene hybrid as a high-capacity anode material for lithium ion batteries. J Am Chem Soc 132:13978–13980
87. Yang S, Feng X, Ivanovici S, Müllen K (2010) Fabrication of graphene-encapsulated oxide nanoparticles: towards high-performance anode materials for lithium storage. Angew Chem Int Ed Engl 49:8408–8411
88. Zhou G, Wang D-W, Li F, Zhang L, Li N, Wu Z-S, Wen L, Lu GQ, Cheng H-M (2010) Graphene-wrapped $Fe_3O_4$ anode material with improved reversible capacity and cyclic stability for lithium ion batteries. Chem Mater 22:5306–5313
89. Wu Z-S, Ren W, Wen L, Gao L, Zhao J, Chen Z, Zhou G, Li F, Cheng H-M (2010) Graphene anchored with $Co_3O_4$ nanoparticles as anode of lithium ion batteries with enhanced reversible capacity and cyclic performance. ACS Nano 4:3187–3194
90. Wang JZ, Zhong C, Wexler D, Idris NH, Wang ZX, Chen LQ, Liu HK (2011) Graphene-encapsulated $Fe_3O_4$ nanoparticles with 3D laminated structure as superior anode in lithium ion batteries. Chemistry 17:661–667
91. Zhang L-S, Jiang L-Y, Yan H-J, Wang WD, Wang W, Song W-G, Guo Y-G, Wan L-J (2010) Mono dispersed $SnO_2$ nanoparticles on both sides of single layer graphene sheets as anode materials in Li-ion batteries. J Mater Chem 20:5462–5467
92. Li B, Cao H, Shao J, Li G, Qu M, Yin G (2011) $Co_3O_4$@ graphene composites as anode materials for high-performance lithium ion batteries. Inorg Chem 50:1628–1632
93. Xia G, Li N, Li D, Liu R, Wang C, Li Q, Lu X, Spendelow J, Zhang J, Wu G (2013) Graphene/$Fe_2O_3$/$SnO_2$/ternary nanocomposites as a high-performance anode for lithium ion batteries. ACS Appl Mater Interfaces 5:8607–8614
94. Hummers WS, Offeman RE (1958) Preparation of Graphitic Oxide. J Am Chem Soc 80:1339
95. Marcano DC, Kosynkin DV, Berlin JM, Sinitskii A, Sun Z, Slesarev A, Alemany LB, Lu W, Tour JM (2010) Improved Synthesis of Graphene Oxide. ACS Nano 4:4806–4814
96. Ramesha GK, Sampath S (2009) Electrochemical reduction of oriented graphene oxide films: an in situ Raman spectroelectrochemical study. J Phys Chem C 113:7985–7989

# Chapter 5
# Graphene Oxides in Filtration and Separation Applications

**Zhiping Xu**

**Abstract** Graphene oxides feature much richer structural and physiochemical properties than the two-dimensional crystal graphene they are derived from. The stacked structures as well as functional groups and defects in the monatomic layers lead to a porous microstructure and engineerable channels for selective transport of water, ions, and gases across the graphene oxide membranes. Additional merits include their facile fabrication, low cost, and flexibility. Recent efforts in exploring the structure–property relationship of these materials and environment and energy-related applications not only demonstrate excellent balance between the permeability and selectivity of fluid transport through the graphene oxide membranes, but also deepen our understanding of the molecular transport mechanisms down to the nanoscale. In this chapter we review some of the major theoretical and experimental advances in this field, along with our perspectives for the future development.

**Keywords** Graphene oxides • Filtration • Separation • Microstructure • Permeation • Selectivity

## 5.1 Introduction to Molecular Filtration and Separation

Recent crisis of water resource shortage, especially in developing countries, has raised immersed research interests in new clean water technologies [1]. As illustrated in Fig. 5.1, membrane-based technologies such as ultrafiltration (UF), nanofiltration (NF), reverse osmosis (RO), and forward osmosis (FO) have received enormous attention because of their low cost and effective energy utilization for removing contaminants from water. These technologies have already been widely used in a broad spectrum of industries such as water purification and food processing. Designed for membrane-based water desalination and wastewater treatment applications, porous

Z. Xu (✉)
Applied Mechanics Laboratory, Department of Engineering Mechanics
and Center for Nano and Micro Mechanics, Tsinghua University, Beijing 100084, China
e-mail: xuzp@tsinghua.edu.cn

© Springer International Publishing Switzerland 2015
W. Gao (ed.), *Graphene Oxide*, DOI 10.1007/978-3-319-15500-5_5

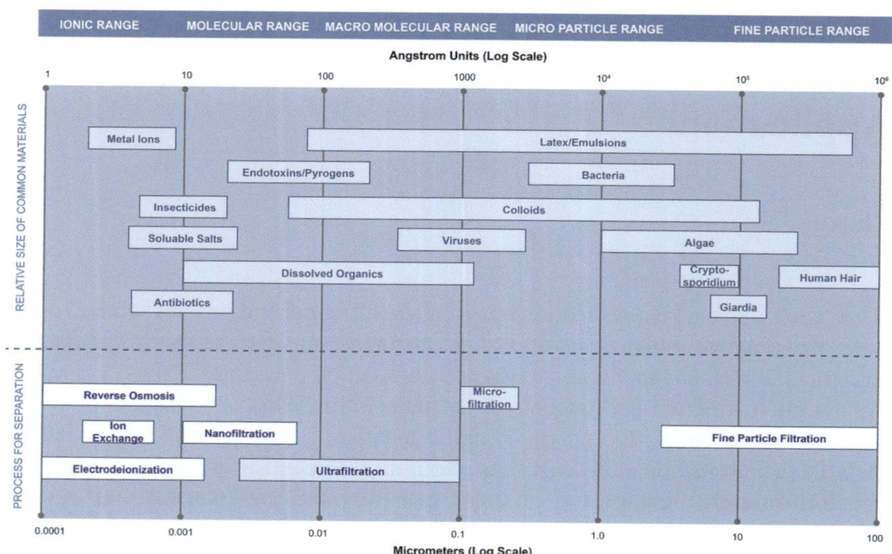



**Fig. 5.1** Filtration and separation spectrum that shows characteristic length scales and mechanisms of the technologies

nanostructures with tubular or laminar topologies have been explored by conducting precise measurements in experiments and molecular level computer simulations [2–6]. Using porous structures with a dimension on the order of one nanometer that is close to the size of water molecules or ions, one could achieve high salt rejection in water desalination or high selectivity for ion separation. Moreover, researches have demonstrated unexpected high permeability of water through nanochannels due to the fact that the flow resistance is remarkably reduced under nanoscale constriction [7–9]. As a result, with an optimally designed architecture, nanostructured membranes could be scaled up for industrial applications with performance far beyond existing technologies.

Compared to conventional membrane materials such as the zeolites, porous silica, and polymers, the graphene oxide (GO) membrane with functionalized monatomic layers as building blocks is an ideal material for these applications. The rich chemistry of functional groups results in differentiation in the interaction between ions and GO. Its tunable microstructures and physiochemical properties thus allow rational design of selective fluidic transport at the molecular level, offering exciting opportunities in establishing high-performance and low-cost clean water technologies. GO membranes have also been considered as permeable membranes for selective gas transport. Their nanoporous microstructures and species-specified adsorption enable gas separation or capture based on the mechanisms of molecular sieving or adsorption. Promising applications could be found in the environmental and energy industry, e.g., $CO_2$ capture from flue gas, $CO_2$ removal from natural gases, $CO_2$ recovery from landfill gas, and selective hydrogen separation.

In the following part of this chapter, we first introduce the atomic structures of GO and microstructures of GO membranes in Sect. 5.2, and the mechanism of filtration and separation in Sect. 5.3. A few representative works in developing relevant applications that have been reported recently are reviewed in Sect. 5.4, followed by our concluding remarks and perspectives in Sect. 5.5.

## 5.2  Structures of GO and GO Membranes

As discussed previously in Chap. 1, the GO layer consists of oxygen-rich functional groups as well as vacancy-like defects within the sheet. The density and spatial distribution of these imperfections, as well as their modulation of the microstructures of GO membranes, define the strength of gas adsorption, resistance to molecular flow, and ionic affinity in various environments. From a design point of view, as the hydrophilicity and solubility of GO sheets allow them to be well dispersed in solution as single-layer sheets, their atomic structures could then be engineered before assembled into membranes. In this section we review briefly current understandings of the atomic and microscale structures of GO membranes, before continuing discussions on their implications in the separation and filtration processes.

### 5.2.1  Atomic Structures of Graphene Oxide

*Models of graphene oxides*: As a product of strong acid/base treatment, the atomic structure of GO is quite disordered compared to the hexagonal lattice of crystalline graphene. The single-layer sheet usually contains oxygen-rich functional groups such as hydroxyl and epoxide groups on the carbon basal plane, carboxyl, carbonyl, and phenol groups to the edges of sheets or nanoholes embedded, as well as lactol rings [10, 11]. A schematic illustration is shown in Fig. 5.2a. These functional groups are usually negatively charged and interact with molecules or ions with different strengths according to their atomic charges and polarization. The covalently bonded network in GO is defective after the oxidation-reduction treatment. Experimental evidence studies suggest the existence of defect-free graphene areas with size of a few nanometers interspersed with defect areas that are dominated by clustered pentagons and heptagons (Fig. 5.2b). All carbon atoms in the defective areas are bonded to three neighbors maintaining a planar $sp^2$ configuration with relative chemical stability. However, significant in-plane distortions and strain in the surrounding lattice as well as out-of-plane displacement are introduced [12]. Considering that interlayer gallery, interedge spaces, and embedded pores can be occupied by the fluid, both basal-plane and edge functionalization play important roles in determining the percolated paths for molecular transport, and thus the permeability and selectivity in separation and filtration applications (Fig. 5.2c).

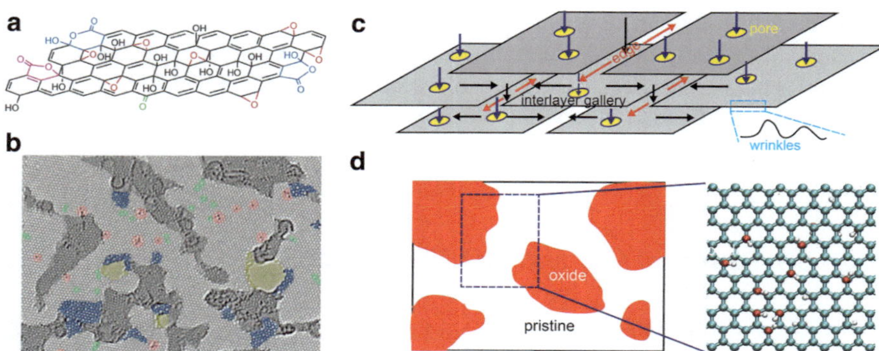



**Fig 5.2** (**a**) Atomic structures of a single-layer GO sheet, taking into account the five- and six-membered lactol rings, ester of a tertiary alcohol, hydroxyl, epoxy, and ketone functionalities. The model here only shows the chemical connectivity, but not the steric orientation of these functionalities [11]. (**b**) Defective structures of GO characterized by the high-resolution transmission electron microscopy. The defect-free crystalline graphene area is displayed in *light-gray color*. Contaminated regions are shaded in *dark gray*. *Blue regions* are the disordered single-layer carbon networks, or extended topological defects, that are identified as remnants of the oxidation–reduction process. *Red areas* highlight individual ad-atoms or substitutions. *Green areas* indicate isolated topological defects, that is, single-bond rotations or dislocation cores. Holes and their edge reconstructions are colored in *yellow*. The scale bar is 1 nm [66]. (**c**) Illustrated microstructures of GO membranes that comprise stacked GO sheets. Paths opened for molecular transport consist of the interlayer gallery, slit-like interedge spacings, embedded pores, and channels formed by the wrinkles of GO sheets [5, 6]. (**d**) An atomistic model of GO constructed for atomistic simulations [5, 6]. (Reproduced by permission from NPG, ACS)

*Clustering of functional groups*: In addition to the type of functional groups and defects, their spatial distribution is also important in specifying the percolated path for molecular transport. Results from experimental characterization and computer simulations have been reported to support that the basal-plane imperfections tend to agglomerate and form clusters, leading to a patched nature of structural modification comprising both pristine and oxide regions, as well as nanoholes (Fig. 5.2d) [12–15]. As a result, a 2D percolated path along the unfunctionalized region could be formed in the interlayer gallery to enable fluidic transport with low wall friction.

*Surface modification and interlayer intercalation*: One of the most attractive features of GO is that the structure could be functionalized or modified before the isolated sheets are assembled into membranes. This process could be designed to tune the density and spatial distribution of functional groups, defects, the nanopores they induce, as well as the surface charge and lateral size of GO sheets [16–20]. The structure of GO membranes in liquid solvents is found to be quite dependent on the liquid molecules. While GO membranes are easily hydrated by water, intercalation of alcohols is hindered and the insertion of ethanol into the membrane is limited to only one monolayer [21]. For GO membranes in contact with solution or humid air, the interlayer gallery changes upon hydration. The negatively charged, highly hydrophilic surface of GO favors the intercalation of water molecules that could lead to the formation of hydrogen bond (H-bond) networks [22]. The water contact angle of graphene $\theta_{c,G}$ is in the neutral range of 87–127°. However, after functionalization the water contact angle of GO $\theta_{c,GO}$ decreases with the concentration of oxygen-rich functional groups on the sheet. For a typical concentration of O:C$=20$ % for GO, $\theta_{c,GO}$ is calculated to be 26.8° [23]. The hydrophilicity of GO allows the membrane to expand by intercalating water molecules. Experimental measurements show that the interlayer distance measured for bilayered GO increases by ~0.1 nm, from ~0.7 to 0.8 nm, when exposed in the environment with a relative humidity in the range of 2–80 %, while immersion into liquid water raises the value up to ~1.2 nm [24]. According to the patched nature of GO sheet, the functionalized regions act as spacers that keep adjacent crystallites apart, with tunable interlayer distance depending upon the humidity, and prevent them from being dissolved, whereas pristine regions provide a network of capillaries that allow low-friction molecular transport [5, 6, 25, 26]. In addition to water, GO sheets could also be engineered by other functionalization approaches such as ion and polymer intercalation, which control the size and charge characteristics of the transport channels in separation and filtration applications [27–29].

## 5.2.2   Microstructures of Multilayer Graphene Oxide Membranes

For the permeation and selectivity of a porous membrane, not only the physiochemical properties of pore surfaces are relevant, but also the structural characteristics are of great importance in defining the membrane performance, which can be quantified through the porosity $p$ and tortuosity $\tau$. The porosity is the relative volume of the membrane that could be occupied by the fluid, while the tortuosity is the ratio between the average length of the fluid paths and the geometrical length of the sample. The permeability measuring the ratio between the flow rate and pressure gradient could be enhanced by increasing the factor $p/\tau$. The microstructures of GO membranes, although difficult to be resolved in detail experimentally, could be controlled during the fabrication process through, for example, the sheet size, functionalization level, as well as template-enabled controls.

*Brick-and-mortar microstructures*: As illustrated in Fig. 5.2c, the open network for fluid flow across the GO membrane consists of two major paths. The first one travels through the nanopores within GO sheets or the slit-like interedge spacings. Due to the random stacking order during the synthesis, the alignment between these structures embedded in the neighboring layers of the brick-and-mortar structure will only be possible for ultrathin membranes [30]. As the thickness increases, the species in transport must flow through the interlayer gallery before they find a pore or slit to go through. This additional cost in general reduces the flow rate and becomes the rate-limiting process of the cross-membrane flow. The second path for molecular flow is within the interlayer gallery, along the most efficient channels that could be the pristine graphene regions with less resistance against the flow or the wrinkles and nanochannels as we discuss below [25, 31, 32].

Corresponding to these two regimes of molecular transport, two types of membrane could be designed for filtration and separation applications, where the fluid flow is driven across or along the GO sheets. In either case, enlarging the interlayer distance by molecular intercalation or other spacers, increasing the in-plane defects by exotic treatments such as heating, irradiation, or chemical functionalization, or choosing GO sheets with smaller lateral sizes in the fabrication could improve their permeability, although it should be kept in mind that the selectivity and mechanical resistance of membrane could be sacrificed with these modifications.

*Wrinkles*: As a 2D material, the long-range crystalline disorder of graphene could not be maintained in a freestanding manner. Out-of-plane corrugations induced by thermal fluctuation of the presence of defects within the GO sheets become prominent even when they are constrained, unless the graphene or GO sheets are very strongly adhered to a substrate or inside a matrix. The athermal fluctuation of 2D sheets is an intrinsic feature of GO as a defective monatomic layer, arising from the formation of $sp^3$ hybridized atomic structures and topological defects such as non-hexagonal rings. The local structural corrugation thus induced is on the order of one nanometer [12]. These defects and thermal fluctuation are also expected to affect the overall morphology of the sheet and the long-range correlation can hardly be preserved during the membrane synthesis process. The functional groups on GO could roughen the potential surface that describes the interaction between neighboring GO sheets, leading to local stacking structures that lock channel-like structures in between [33]. These microstructural corrugations can become part of the transport channel network in the membrane and could be altered in the solvated and dry conditions as the capillary forces occurring during the drying process could flatten the GO sheets [34]. Tuning the synthesis and fabrication processes of GO is demonstrated to be able to control the microstructures of GO membrane. For example, by changing the temperature of hydrothermal reduction process from 90 to 150 °C, the effective pore size for nanoparticle rejection increases from 3 to 13 nm, which is mainly attributed to the wrinkled membrane structures. Further experimental evidences from direct yellow nanofiltration tests suggest that this control of corrugation could even reach the scale of sub-nanometer [31].

*Nanochannels*: Microstructures of GO membranes could also be rationally engineered to enable or enhance their functions. For example, nanostrand-templated GO membranes display excellent ultrafiltration performance (Fig. 5.3). A network of nanochannels with a narrow-size distribution between 3 and 5 nm is created in the membrane that offers superior separation performance. The permeance shows a tenfold enhancement without sacrificing the rejection rate compared with that of common GO membranes, and is more than 100 times higher than that of commercial ultrafiltration membranes with similar rejection. With a viscous nature of flow, the enhancement is attributed to the enriched porous structures by nanochannels that significantly reduces the overall channel length and enhances the $p/\tau$ ratio. Moreover, an abnormal pressure-dependent separation behavior is identified as the nanochannels could deform reversibly, leading to modulations in the water flux and rejection rate. As a result, the membrane performance can be tuned through pressure control, enabling promising mechanomutable applications.

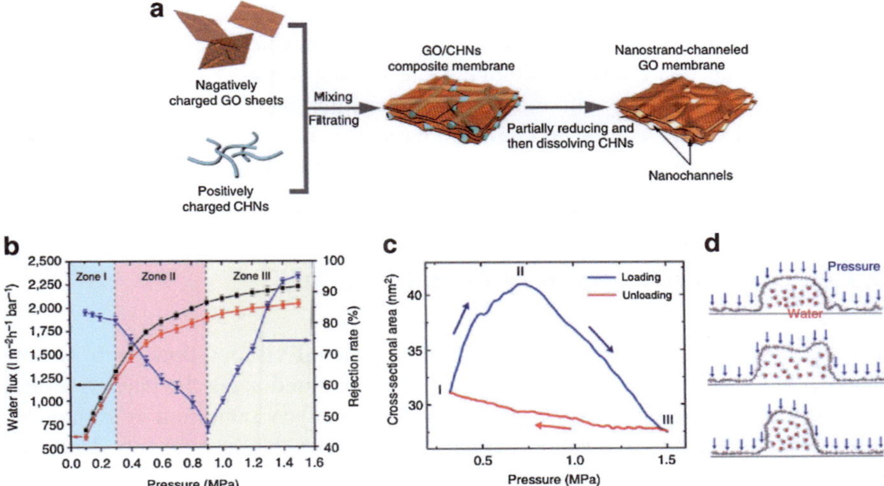


Fig. 5.3 (a) Illustration of the multistep fabrication process of nanostrand-channeled GO membrane, consisting of the formation of a dispersion of positively charged copper hydroxide nanostrands (CHNs) and negatively charged GO sheets on a porous support, followed by hydrazine reduction, and finally nanostrand removal. (b) Pressure-dependent flux and rejection of Evans blue molecules of the membrane. The *black squares* and *red circles* represent the water flux during the first and third pressure-loading cycles, respectively. The *blue triangles* denote the rejection rate within the first cycle. (c) Simulated changes in the cross-sectional area of nanochannel as a function of the applied pressure. (d) The response in deformation of a GO nanochannel under pressure loading, modeled in molecular dynamics simulations. The amplitude of applied pressure increases from top to the bottom. The cross section of nanochannels is flattened first with a larger area and then compressed at higher pressure [32]. (Reproduced by permission from NPG)

## 5.3  Mechanisms of Selective Molecular Transport in Graphene Oxide Membranes

Membrane filtration is usually understood as a dynamic process where fluid dynamics and electrochemistry govern. However, down to the nanometer scale, the conventional theories need to be updated, if not changed, to describe and predict new phenomena in related applications [5, 6, 18–20, 25, 35, 36]. The brick-and-mortar hierarchy of GO membrane allows both cross and in-plane flow of molecules. The first process relies on the presence of vacancy-like defects in GO or slit-like interedge spaces between neighboring GO sheets aligned in plane. Although in cross flow, molecular transport directly across the GO sheets travels much shorter distance than the in-plane flow between neighboring GO layers, its participation is only significant in ultrathin membranes that consist few GO layers, for example, considering GO membranes with defects or slit-like interedge spacings with widths of $w$ and areal densities of $n$. The probability to find a straight path across an $N$-layer membrane then decreases drastically as $2(wn)^{N-1}$ [5, 6]. To enhance the cross-flow permeability, one must increase either the size of pores in GO or their concentration, which will, however, reduce the selectivity and mechanical resistance as a cost. Otherwise, molecular transport within the interlayer spacings is unavoidable and the whole transport network will include both the interlayer gallery and nanopores inside the GO sheet. As a result, confined molecular transport through these nanoporous structures becomes the key to understand the filtration process.

### 5.3.1 Selective Transport of Liquid and Gas

*Viscous flow with slip boundary conditions*: Confined viscous flow such as flow inside a pipe or between parallel plates could be captured using the Navier–Stokes equation in the continuum regime. A non-slip boundary condition is commonly assumed in exploring viscous flow at macroscale, where the flow velocity decreases to zero at the wall surface. However, within nanoscale channels, the atomic structures of fluid tend to deviate from the bulk. Atomistic simulation results show that the intercalated water between GO sheets displays layered structures, from

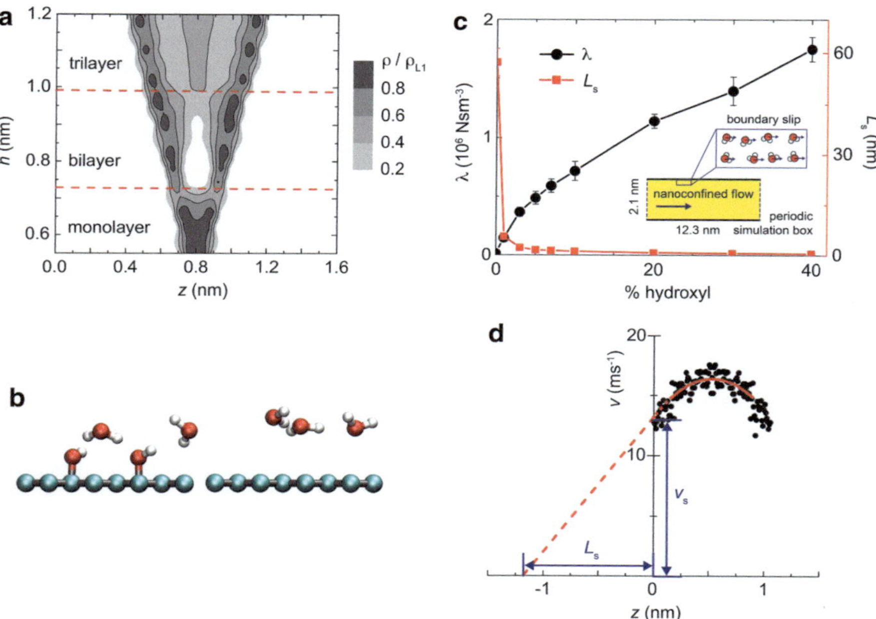

**Fig. 5.4** (**a**) Density profiles of water molecules intercalated between GO sheets, which shows the transition from single layer, bilayer, and trilayer to more disordered structures as the interlayer distance increases. (**b**) Atomistic simulation snapshots that show water molecules in the first water layer at the GO (*left*) and graphene (*right*) wall. The distances between the water molecules and the basal planes are close in these two different structures. (**c**) The dependence of frictional coefficient and interfacial slip length on the concentration of hydroxyl groups functionalized on GO. (**d**) The velocity profile of water flow between GO sheets and the definition of slip length [5, 6]. (Reproduced by permission from APS)

monolayer, bilayer, to trilayer with the interlayer distance below 1.2 nm as illustrated in Fig. 5.4a [5, 6]. Interestingly, although the GO sheet contains surface functional groups, the distance between the first water layer and the basal plane in GO remains the same as that between graphene sheets at the interlayer distance (Fig. 5.4b). As a result, the interlayer gallery in GO membrane is actually a 2D planar channel containing layered water content, with uniform thickness but heterogeneous surface properties if the wrinkles are not considered.

Water transport in nanoscale channels has received much interest recently from both fundamental research and practical applications. Research has shown that the combined effect from nanoconfinement and atomistically smooth graphitic walls leads to significant flow enhancement inside carbon nanotubes (CNTs) or between graphene sheets, in comparison to theoretical predictions from the Navier–Stokes equation supplied with non-slip boundary conditions. The flow enhancement could be quantified through the definition of Navier slip length $L_s = v_s/(dv/dz)|_{z=0}$, where $v_s$ is the slip velocity at the fluid-wall interface and $dv/dz$ is the tangent of the velocity profile along the normal direction $z$ at the wall ($z=0$) (Fig. 5.4c, d). Experimental studies have revealed that for CNTs with an inner pore size of a few nanometers, a slip length over micrometers could be established [37–39]. Although the reported values of $L_s$ have some discrepancy due to the differences in experimental samples, conditions of measurements, or theoretical models, the large boundary slip that elevates the overall flow rate clearly demonstrates the advantage of using nanostructured materials in environmental applications.

Graphene, considered as the zero-curvature limit of the CNT, can also form nanoscale flow channels while stacked into multilayers. However, the van der Waals cohesion between pristine graphene sheets results in an interlayer distance of 0.34 nm that cannot accommodate even a single water molecule. Functionalized graphene sheets, such as GO and polymer-intercalated graphene sheets, offer a wider interlayer space for molecular transport, and the size can be further tuned by engineering the structure or environmental conditions. For non-slip planar flow between GO sheets, the volumetric flux $Q$ can be estimated as $Q_{nonslip} = -d^3 \Delta P W/(12\eta L)$, where $d$ is the channel thickness that should be redefined here as the hydrodynamic thickness by excluding the vacuum space $\delta = 0.5$ nm between water molecules and the carbon basal plane. However, once the boundary slip is considered through the slip length $L_s$, the flow rate could be enhanced by a factor of $\varepsilon = Q_{slip}/Q_{nonslip} = 1 + 6L_s/d$, which is significant if the atomistic smoothness of the wall is well preserved and a large slip length of micrometers is available.

Once the atomistic smoothness is broken by the presence of functional groups as in the situation of GO, the flow enhancement is weakened remarkably. Studies using atomistic simulations show that the enhancement factor could be reduced by two orders, from 48.13 to 0.44 nm for a typical concentration of oxidization groups with O:C = 30 %. That is to say, even with the nanoconfinement, the water flow between GO layers is viscous and the flow rate should be estimated using non-slip boundary conditions [5, 6, 32, 40].

It should be noted that there are some evidences showing that the oxidization of GO sheets is not spatially uniform. Functional groups could group into clustered patterns and leave flow paths comprising pristine graphene channels in the GO membrane. A fast lane for water transport may exist based on the fact that water flow over graphene sheet experiences ultralow friction [25, 41]. However, the oxidized region of GO is also wetted with negligible boundary slip for the flow. Even if a percolated path could be formed along the pristine region of GO, the internal friction between water molecules in the pristine and oxidized regions at their interface could impede flow within the pristine paths. This side-pinning effect can only be neglected if the width of pristine channels is much larger than the size of 1–2 nm that is characterized in experiments [5, 6].

Water transport through nanopores or interedge spaces inside GO was also explored by molecular dynamics simulations, which suggest that the permeability is close to that of flow between GO sheets, but is significantly reduced compared to that between pristine graphene sheets with the same channel width. To estimate the overall permeability across the GO membrane in practical situation, the entrance/exit loss and capillary-driven force must also be considered [25, 42].

*Gas permeation*: Similar to the mass transport of liquid molecules, selective gas permeation through GO membranes could also be established by balancing the performance trade-off between mobility of gas molecules and the selectivity, which could be enabled by either the molecular size-sieving mechanism or the difference in gas adsorption and diffusion behaviors. For gas transport confined by nanoscale pores, the flow regime changes from Knudsen diffusion for mesopores from 2 to 50 nm to constrained molecular diffusion for micropores below 2 nm. The Knudsen flow model fits the situation where the mean free path of gas molecules is longer than the size of pores, where diffusion is limited by thermalized collisions with the wall. Based on this assumption, the permeability could be estimated as $pd/3\tau h$ $(8/\pi mRT)^{1/2}$, where $d$ is the pore size, $m$ is the molecular weight, $R$ is the ideal gas constant, and $T$ is the temperature. Thus the separation of gases could be achieved through differentiating the molecular weight [16, 35].

For the constrained diffusion of gases, the gas permeation is dominantly determined by their interaction with the pore wall and their kinetic diameters. One way to quantify these effects is through the self- and collective or transport diffusivities [43]. Similarly, rapid transport of gases in carbon nanotubes with pristine graphitic walls was reported, which are orders of magnitude faster than in the zeolites with comparable pore sizes [35, 44]. However, in corrugated GO membranes with negatively charged functional groups on the surfaces, the gas permeability would be significantly reduced to the flow between graphene sheets with the same interlayer distance. Moreover, in addition to the selectivity based on molecular weights or kinetic diameters of the gases, interaction with the wall adds another dimension in establishing the separation of polar molecules such as $CO_2$.

### 5.3.2 Selective Ion Transport

One of the major targets of water desalination, recovery, and recycling processes is to selectively remove ions from the water. The ionic selectivity is defined by the size, shape of ions, as well as their interaction with the hydrophilic, negatively charged GO membrane. In aqueous form, ions are gel-like by possessing

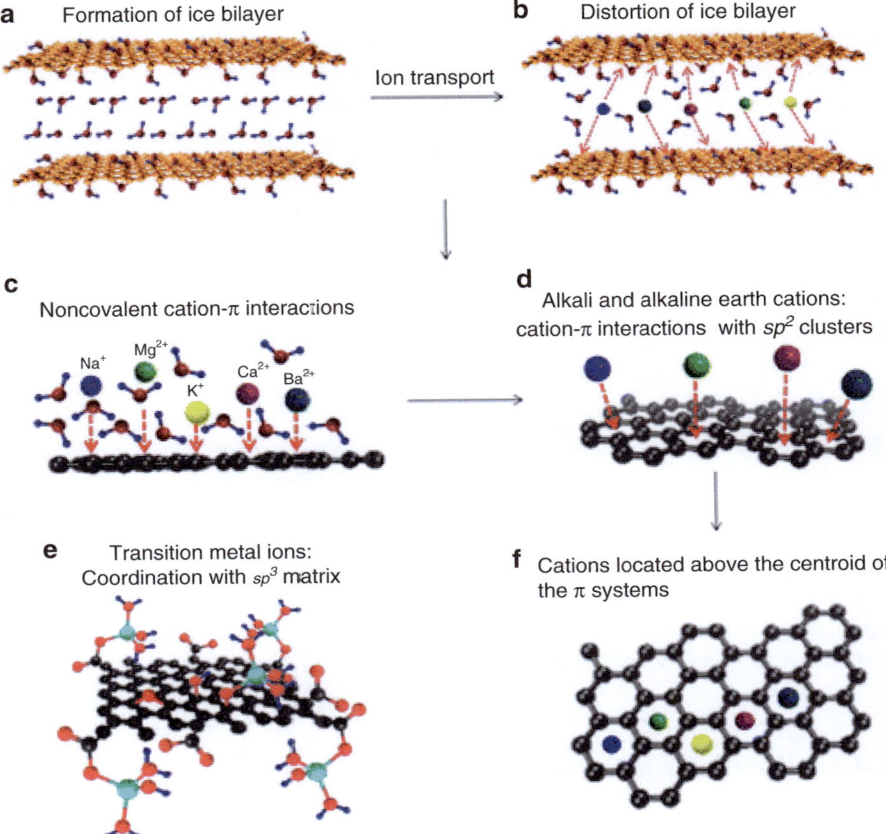



**Fig. 5.5** Schematic diagrams of selective ion transport through GO membranes and the diverse interactions between different cations and GO [18–20]. (Reproduced by permission from ACS)

surrounding hydration shells of water molecules. The strength of hydration depends on the type of ions. For example, small monovalent ions ($Li^+$, $Na^+$, $F^-$) feature much stronger hydration energy than larger ones ($K^+$, $Cl^-$, $Br^-$, $I^-$). The ions possessing higher charge densities also bind stronger with larger water clusters. Moreover, the anions hold their hydration shells relatively more strongly than the cations due to the difference in the strength of ion-water interaction [45] (Fig. 5.5).

For ions confined between nanochannels with pore sizes comparable to those of the hydrated ions, such as those that conduct across the GO membranes, the solvation shell could be strongly disturbed. Considering the similarity between negatively charged oxygen-rich groups on the GO surface and polar water molecules, the mechanisms of ionic permeation and selection become more complicated than the simple picture of molecular sieving. Firstly, the interaction between GO and the ions could then differ by the level of ion hydration, and the oxygen-rich groups could behave as part of the solvation shell. At the same interlayer distance, electrostatic interaction between GO and small ions could be better screened and the permeation rate could then be higher [18–20]. Secondly, as the width of nanochannels becomes comparable or smaller than the Debye screening length, the interaction between the oxygen-rich groups on GO and conducting species starts to play an important role. As both pristine and oxidized regions coexist in the GO membrane, not only their coordination with transition metal ions but also the cation–$\pi$ interaction between the graphitic regions and main-group cations are responsible for the selective ion transport in addition of sieving the solvated ions by their sizes [18–20, 36, 40].

## 5.4 Recent Progresses in the Experimental Demonstrations Towards Applications

With the capabilities of measuring the water and ion permeance, selectivity, as well as characterizing the microstructures of GO membranes, significant efforts have been made recently to understand the underlying mechanisms of selective liquid and gas transport, and explore their potential applications in the filtration and separation applications. The proposed mechanisms introduced in the previous section are justified based on evidences from experimental and computational studies, to upgrade the current understandings of the ultimate design of membranes for filtration and separation by engineering single-atom-thick 2D materials.

### 5.4.1 Water Purification

Submicrometer-thick GO membranes were reported to be completely impermeable to liquids, vapors, and gases including He, but allow unimpeded permeation of water, which is at least $10^{10}$ times faster than He [25]. The measurements were done by analyzing the evaporation rates of species across GO membranes from their weight loss. Impressively, the unimpeded water permeation in submicrometer GO membranes is as fast as water evaporation through an open aperture, suggesting an evaporation-limited regime. The fast water permeation through the membrane is explained as the presence of a high capillary pressure formed between GO sheets, and a large, percolated path of pristine, unoxidized graphene regions that allow water flow with ultralow friction. Moreover, compared to the as-prepared GO, the interlayer distance in reduced GO by thermal annealing decreases significantly from ~1 to ~0.4 nm. As a result, the water permeability is reduced remarkably by 100 times, which indicates the significance of in-plane transport within the interlayer gallery and the critical role of interlayer distance in the molecular sieving process.

Reduced GO membranes with a thickness of 22–53 nm were fabricated on a microporous substrate for use in nanofiltration-based water purification [30]. These ultrathin membranes demonstrate high pure water flux of 21.8 $Lm^{-2}$ $h^{-1}$ $bar^{-1}$, high retention for organic dyes >99 %, and moderate retention of ~20–60 % for ion salts. Both physical sieving and electrostatic interaction between the charged groups in GO the rejection process and analysis of the flow rate suggest the possibilities of water flow through nanopores and slits within the GO layers. It is further shown that as the membrane thickness increases from 22 to 53 nm, the pure water flux across the membrane decreases dramatically from 21.8 to 3.26 Lm$^{-2}$ h$^{-1}$ bar$^{-1}$. Moreover, the retention measured for different salt solutions agrees with theoretical predictions from the Donnan exclusion theory, which links the salt retention to the valences of cationic and anionic species.

## 5.4.2 Ion Separation

Ion transport through GO membranes was investigated using a nanofluidic device [46]. The massive parallel 2D channels between neighboring GO layers are percolated that allow efficient in-plane ionic transport. The ionic conductance measured is as high as that of the bulk electrolyte solution at high salt concentration. However, at low salt concentration below ~0.05 M, the conductance deviates from bulk behavior and displays a concentration-independent, surface-charge-governed regime of ionic transport (Fig. 5.6a–d). The negatively charged nature of GO also leads to high affinity to the cations while anions are repelled, allowing ion separation. The flexibility of the GO paper-based nanofluid device was also demonstrated by mechanical folding for 30 cycles.

There are also experimental evidences showing that the hydrated radius of salt ions in a molecular sieving regime determines the rate of ion permeation [26]. The ions with sizes below the critical mesh size of the nanochannel permeate at almost the same rate, whereas large ions and organic molecules exhibit no detectable permeation (Fig. 5.6e). These results suggest that, to avoid ionic conduction, the size of interlayer channel should be controlled down at ~0.6 nm to allow monolayer water

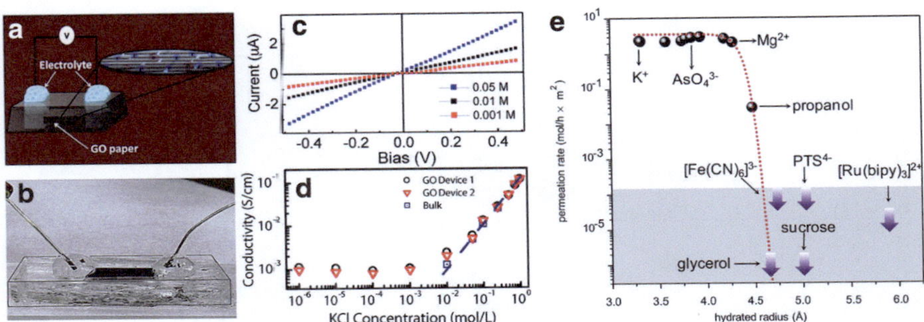


**Fig. 5.6** (**a**) Schematic illustration and (**b**) photograph showing GO paper-based nanofluidics device. (**c**) Representative *I–V* curves recorded at different KCl concentrations. (**d**) Ionic conductivity as a function of salt concentration measured through GO nanochannels [46]. (**e**) Ionic permeation rates measured by using 5-µm-thick GO membranes. The *thick arrows* are detection limit in the measurements, and the *dashed curve* is a guide to the eye, showing an exponentially sharp cutoff at 0.45 nm with a width of ~0.01 nm [26]. (Reproduced by permission from ACS, AAAS)

transport and reject even the smallest salts. Nanopores or interedge slits inside GO are expected to have the same critical size for ionic transport. However, a recent work shows that, as the extreme case of ionic transport, thermal protons are able to permeate through even perfect graphene lattice with a surprisingly low resistivity of $10^{-3}$ $\Omega$cm$^2$ above 250 °C. The transport can be further enhanced by decorating the graphene monolayer with catalytic metal nanoparticles that could lower the tunnel barrier [47].

Selective ion permeation across GO membranes was studied by considering a wide spectrum of ion types. It was reported that sodium ions permeate much faster than heavy-metal salts such as $Mn^{2+}$, $Cd^{2+}$, and $Cu^{2+}$. Notably, copper sulfate and rhodamin B are blocked entirely because of their strong interactions with GO [36, 40, 48]. Specifically, the transition metal ions could be coordinated by the oxygen-rich groups on GO and thus be strongly impeded. Further investigation on alkali and alkaline earth cations shows relative transport rates in the order of $H^+ > Mg^{2+} > Na^+ > Ba^{2+}$, $Ca^{2+}$, and $K^+$, from the highest value to the lowest ones, which can be explained by a combined effect of $\pi$-cation interaction and nanoconfined hydration [18–20]. The permeability increases as the lateral size of GO sheet is reduced, where a shorter path is available for the ionic conduction. Moreover, a remarkable temperature dependence is observed from 20 to 40 °C. The ion penetrations increase significantly with temperature, indicating the role of thermal activation in the permeation. The ion transport through GO membrane is further demonstrated to be controllable under electromagnetic fields. Specifically, electric field can either increase or decrease the ion transport depending on the flow direction, while magnetic field can enhance the ion penetration monotonically [18–20] (Fig. 5.7).

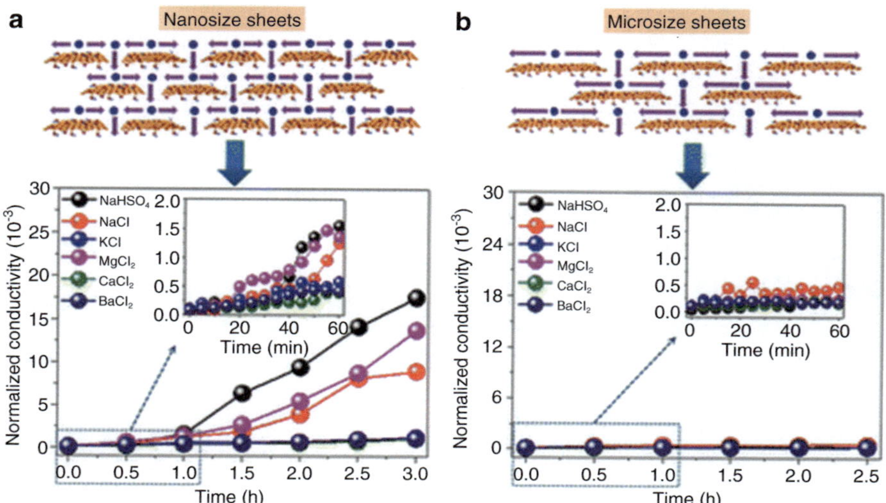


**Fig. 5.7** Schematics of the lamellar structures and conductivities of cations across GO membranes composed of (**a**) nanosized and (**b**) microsized GO sheets. *Insets* show the initial stage (0–60 min) measurements of the penetrations [18–20]. (Reproduced by permission from ACS)

### 5.4.3  Gas Separation

A few experimental researches have been reported recently to explore selective gas transport across graphene (with in-plane defects generated during the growth and transferring processes) and GO membranes. Few and several layered membranes were engineered to exhibit excellent gas separation characteristics (Fig. 5.8). The close-packed, interlocked GO layered structures display an explicit effect on the selective gas transport. At 298 K, the order of gas permeability is $CO_2 > H_2 > He > CH_4 > O_2 > N_2$. The gas transport is determined by the pore size and the molecular mean free path in a gas mixture, and as in the Knudsen diffusion regime, gases are separated based on their differences in the molecular weight. The observed $CO_2$-philic behavior of permeation can be further enhanced by the presence of water, and high $CO_2/N_2$ selectivity can be achieved by well-interlocked GO membranes in a high relative humidity. More in-plane porous structures could be created by heat treatment, which leads to more permeable selective performance [16].

Gas-separation GO membranes down to 1.8 nm in thickness show $H_2/CO_2$ and $H_2/N_2$ selectivities of mixture separation that are one to two orders of magnitude higher than those of the state-of-the-art microporous membranes. The measured permeability is related to the kinetic diameter of gas molecules. Hydrogen molecules with a kinetic diameter of 0.289 nm permeate ~300 times faster than $CO_2$ (0.33 nm) through an 18-nm-thick GO membrane at 20 °C. The slight difference between their kinetic diameters suggests that a critical mesh size of the pore is ~0.289–0.33 nm. $O_2$ and $N_2$ show similar permeability as $CO_2$, while CO and $CH_4$ have slightly higher permeability although their kinetic diameters are larger. The membrane shows mixture

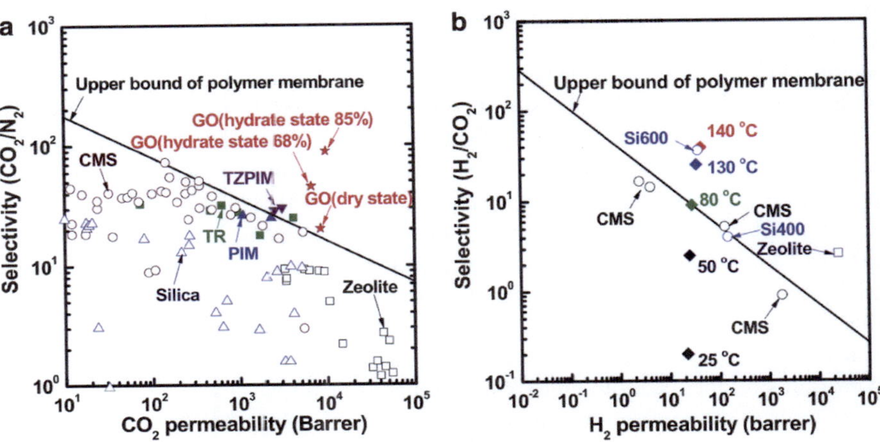



**Fig. 5.8** (**a**) Relation between $CO_2$ permeability and $CO_2/N_2$ selectivity of GO membranes under dry and humidified conditions. (**b**) Comparison of gas separation performance between GO membranes and other membranes for the $H_2$ permeability and $H_2/CO_2$ selectivity [16]. (Reproduced by permission from AAAS)

separation selectivities as high as 3,400 and 900 for $H_2/O_2$ and $H_2/N_2$ through structural defects in GO, which are one to two orders of magnitude higher than those of the state-of-the-art microporous membranes. Moreover, the gas permeabilities decrease exponentially on the membrane thickness, which arises from the changes in the length of molecular transport paths [49].

Another detailed study shows that the intercalation of gas molecules is highly affected by the affinity between hydrophilic surface of GO and target molecules including $CO_2$, $CH_4$, $H_2$, and $N_2$. Only $CO_2$ can be intercalated into the GO membrane. However, the swelling of interlayer spaces in water changes the intercalation phenomena significantly. All of the tested gases could then be intercalated, but the $CO_2$ could be mostly solvated, surrounded by the intercalated water molecules and partially with functional groups on GO surfaces. Moreover, because the interaction between the intercalated water and GO retards the dynamics of water molecules, enhanced gas storage could be achieved within the intercalated water [50].

## 5.5 Conclusive Remarks and Perspectives

The understanding of nanoscale molecular transport mechanisms is of key importance to develop high-performance filtration and separation membranes. However, the current status is far from being mature. For transport through GO membranes, quantitative comparison between theoretical predictions and experimental measurements can hardly be made without detailed information of the microstructures. For example, the length of water transport channel can only be estimated very roughly based on the membrane thickness and average interlayer distance between neighboring GO sheets [25, 30, 32, 40]. To solve this issue, single-channel measurements should be carried out [51, 52], or the membrane fabricated should have a well-defined stacking order, uniform distribution of sheet size, shape, and functionalization of the GO sheets. Computational studies provide some insights into the process [5, 6], but the practical situation could be much different and complicated due to factors including the pH, salt concentration, pressure, thermal fluctuation, and electromagnetic fields, to list a few, which are difficult to be considered through atomistic models and are more appropriate to be explored experimentally [18–20, 53, 54].

Although impressive performance of GO membranes has been demonstrated by lab tests, there is still a gap existing towards their commercial utilization in industrial filtration and separation. To this end, large-scale fabrication of GO membranes that can maintain their structural stability in harsh environment should be established. Swelling of the GO laminates in solution should be avoided by mechanical constraints or chemical binding [26]. The mechanical resistance of GO to applied pressure is much weaker compared to the pristine graphene due to the presence of defects and surface functional groups, which should be assured especially for ultra-thin membranes that are preferred for high permeance [55–57]. GO membranes could also be used as coating or barrier materials in related applications, where the supporting porous materials could provide the required structural stability [16, 49,

58]. However, the support layer could significantly weaken the performance of GO membranes for reasons such as the internal concentration polarization (ICP), which inevitably occurs inside the support layer and leads to a dramatic decrease in membrane flux and rejection rate. As a result, free-suspended GO membranes with desired mechanical resistance are highly demanded for industrial applications [59]. Moreover, the oxygen-rich groups could absorb either polar molecules or charged ions, resulting in significant amount of deposited residuals, which could break down the membrane performance and should be removed for regeneration.

There are also promising opportunities along the direction of membrane development using 2D materials, which can be considered as the ultimate material design due to the fact that all the atoms in these materials are exposed to the environment for engineering and the atomic structure could be tailored by a broad class of techniques. Efforts have already been made by considering other monolayer materials such as $MoS_2$ and graphyne [40, 60–63]. The molecular transport process through open channels in these materials could also be coupled with other functional processes such as electricity generation and energy harvesting [18–20, 52, 64]. Understandings of these functional processes could also inspire studies on transport and energetic processes in biological systems [65].

# References


1. Shannon MA, Bohn PW et al (2008) Science and technology for water purification in the coming decades. Nature 452(7185):301–310
2. Kalra A, Garde S et al (2003) Osmotic water transport through carbon nanotube membranes. Proc Natl Acad Sci USA 100(18):10175–10180
3. Raghunathan AV, Aluru NR (2006) Molecular understanding of osmosis in semipermeable membranes. Phys Rev Lett 97(2):024501
4. Cohen-Tanugi D, Grossman JC (2012) Water desalination across nanoporous graphene. Nano Lett 12(7):3602–3608
5. Wei N, Peng X et al (2014) Breakdown of fast water transport in graphene oxides. Phys Rev E 89(1):012113
6. Wei N, Peng X et al (2014) Understanding water permeation in graphene oxide membranes. ACS Appl Mater Interfaces 6(8):5877–5883
7. Joseph S, Aluru NR (2008) Why are carbon nanotubes fast transporters of water? Nano Lett 8(2):452–458
8. Thomas JA, McGaughey AJH (2008) Reassessing fast water transport through carbon nanotubes. Nano Lett 8(9):2788–2793
9. Falk K, Sedlmeier F et al (2010) Molecular origin of fast water transport in carbon nanotube membranes: superlubricity versus curvature dependent friction. Nano Lett 10(10):4067–4073
10. Lerf A, He H et al (1998) Structure of graphite oxide revisited. J Phys Chem B 102(23):4477–4482
11. Gao W, Alemany LB et al (2009) New insights into the structure and reduction of graphite oxide. Nat Chem 1(5):403–408
12. Gómez-Navarro C, Weitz RT et al (2007) Electronic transport properties of individual chemically reduced graphene oxide sheets. Nano Lett 7(11):3499–3503
13. Kim S, Zhou S et al (2012) Room-temperature metastability of multilayer graphene oxide films. Nat Mater 11(6):544–549

14. Kumar PV, Bardhan NM et al (2013) Scalable enhancement of graphene oxide properties by thermally driven phase transformation. Nat Chem 6(2):151–158
15. Zhou S, Bongiorno A (2013) Origin of the chemical and kinetic stability of graphene oxide. Sci Rep 3:2484
16. Kim HW, Yoon HW et al (2013) Selective gas transport through few-layered graphene and graphene oxide membranes. Science 342(6154):91–95
17. Andrikopoulos KS, Bounos G et al (2014) The effect of thermal reduction on the water vapor permeation in graphene oxide membranes. Adv Mater Interf 1(8):1400250
18. Sun P, Zheng F et al (2014) Electro- and magneto-modulated ion transport through graphene oxide membranes. Sci Rep 4:6798
19. Sun P, Zheng F et al (2014) Selective trans-membrane transport of alkali and alkaline earth cations through graphene oxide membranes based on cation–π interactions. ACS Nano 8(1):850–859
20. Sun P, Zheng F et al (2014) Realizing synchronous energy harvesting and ion separation with graphene oxide membranes. Sci Rep 4:5528
21. Talyzin AV, Hausmaninger T et al (2014) The structure of graphene oxide membranes in liquid water, ethanol and water-ethanol mixtures. Nanoscale 6(1):272–281
22. Compton OC, Cranford SW et al (2011) Tuning the mechanical properties of graphene oxide paper and its associated polymer nanocomposites by controlling cooperative intersheet hydrogen bonding. ACS Nano 6(3):2008–2019
23. Wei N, Lv C et al (2014) Wetting of graphene oxide: a molecular dynamics study. Langmuir 30(12):3572–3578
24. Rezania B, Severin N et al (2014) Hydration of bilayered graphene oxide. Nano Lett 14(7): 3993–3998
25. Nair RR, Wu HA et al (2012) Unimpeded permeation of water through helium-leak-tight graphene-based membranes. Science 335(6067):442–444
26. Joshi RK, Carbone P et al (2014) Precise and ultrafast molecular sieving through graphene oxide membranes. Science 343(6172):752–754
27. Park S, Lee K-S et al (2008) Graphene oxide papers modified by divalent ions: enhancing mechanical properties via chemical cross-linking. ACS Nano 2(3):572–578
28. Gao Y, Liu L-Q et al (2011) The effect of interlayer adhesion on the mechanical behaviors of macroscopic graphene oxide papers. ACS Nano 5(3):2134–2141
29. Hu M, Mi B (2013) Enabling graphene oxide nanosheets as water separation membranes. Environ Sci Technol 47(8):3715–3723
30. Han Y, Xu Z et al (2013) Ultrathin graphene nanofiltration membrane for water purification. Adv Funct Mater 23(29):3693–3700
31. Qiu L, Zhang X et al (2011) Controllable corrugation of chemically converted graphene sheets in water and potential application for nanofiltration. Chem Commun 47(20):5810–5812
32. Huang H, Song Z et al (2013) Ultrafast viscous water flow through nanostrand-channelled graphene oxide membranes. Nat Commun 4:3979
33. Xu Z, Buehler MJ (2010) Geometry controls conformation of graphene sheets: membranes, ribbons, and scrolls. ACS Nano 4(7):3869–3876
34. Zhang X, Wang Y et al (2011) Evaporation-induced flattening and self-assembly of chemically converted graphene on a solid surface. Soft Matter 7(19):8745–8748
35. Verweij H, Schillo MC et al (2007) Fast mass transport through carbon nanotube membranes. Small 3(12):1996–2004
36. Sun P, Zhu M et al (2013) Selective ion penetration of graphene oxide membranes. ACS Nano 7(1):428–437
37. Majumder M, Chopra N et al (2005) Nanoscale hydrodynamics: enhanced flow in carbon nanotubes. Nature 438(7064):44
38. Holt JK, Park HG et al (2006) Fast mass transport through sub-2-nanometer carbon nanotubes. Science 312(5776):1034–1037
39. Kannam SK, Todd BD et al (2013) How fast does water flow in carbon nanotubes? J Chem Phys 138(9):094701

40. Sun L, Huang H et al (2013) Laminar $MoS_2$ membranes for molecule separation. Chem Commun 49(91):10718–10720
41. Mi B (2014) Graphene oxide membranes for ionic and molecular sieving. Science 343(6172):740–742
42. Walther JH, Ritos K et al (2013) Barriers to superfast water transport in carbon nanotube membranes. Nano Lett 13(5):1910–1914
43. Skoulidas AI, Sholl DS (2002) Transport diffusivities of $CH_4$, $CF_4$, He, Ne, Ar, Xe, and $SF_6$ in silicalite from atomistic simulations. J Phys Chem B 106(19):5058–5067
44. Skoulidas AI, Ackerman DM et al (2002) Rapid transport of gases in carbon nanotubes. Phys Rev Lett 89(18):185901
45. Tansel B, Sager J et al (2006) Significance of hydrated radius and hydration shells on ionic permeability during nanofiltration in dead end and cross flow modes. Sep Purif Technol 51(1):40–47
46. Raidongia K, Huang J (2012) Nanofluidic ion transport through reconstructed layered materials. J Am Chem Soc 134(40):16528–16531
47. Hu S, Lozada-Hidalgo M et al (2014) Proton transport through one-atom-thick crystals. Nature 516:227–230
48. Xu Z (2013) Mechanics of metal-catecholate complexes: the roles of coordination state and metal types. Sci Rep 3:2914
49. Li H, Song Z et al (2013) Ultrathin, molecular-sieving graphene oxide membranes for selective hydrogen separation. Science 342(6154):95–98
50. Kim D, Kim DW et al (2014) Intercalation of gas molecules in graphene oxide interlayer: the role of water. J Phy Chem C 118(20):11142–11148
51. Duan C, Majumdar A (2010) Anomalous ion transport in 2-nm hydrophilic nanochannels. Nat Nanotechnol 5(12):848–852
52. Siria A, Poncharal P et al (2013) Giant osmotic energy conversion measured in a single trans-membrane boron nitride nanotube. Nature 494(7438):455–458
53. Talyzin AV, Solozhenko VL et al (2008) Colossal pressure-induced lattice expansion of graphite oxide in the presence of water. Angew Chem Int Ed 47(43):8268–8271
54. Huang H, Mao Y et al (2013) Salt concentration, pH and pressure controlled separation of small molecules through lamellar graphene oxide membranes. Chem Commun 49(53):5963–5965
55. Paci JT, Belytschko T et al (2007) Computational studies of the structure, behavior upon heating, and mechanical properties of graphite oxide. J Phy Chem C 111(49):18099–18111
56. Suk JW, Piner RD et al (2010) Mechanical properties of monolayer graphene oxide. ACS Nano 4(11):6557–6564
57. Song Z, Xu Z et al (2013) On the fracture of supported graphene under pressure. J Appl Mech 80(4):040911
58. Gao W, Majumder M et al (2011) Engineered graphite oxide materials for application in water purification. ACS Appl Mater Interf 3(6):1821–1826
59. Liu H, Wang H et al (2014) Facile fabrication of freestanding ultrathin reduced graphene oxide membranes for water purification. Adv Mater 27(2):249–54
60. Lin S, Buehler MJ (2013) Mechanics and molecular filtration performance of graphyne nanoweb membranes for selective water purification. Nanoscale 5(23):11801–11807
61. Xue M, Qiu H et al (2013) Exceptionally fast water desalination at complete salt rejection by pristine graphyne monolayers. Nanotechnology 24(50):505720
62. Zhu C, Li H et al (2013) Quantized water transport: ideal desalination through graphyne-4 membrane. Sci Rep 3:3163
63. Kou J, Zhou X et al (2014) Graphyne as the membrane for water desalination. Nanoscale 6(3):1865–1870
64. Yin J, Li X et al (2014) Generating electricity by moving a droplet of ionic liquid along graphene. Nat Nanotechnol 9(5):378–383
65. Alberts B, Johnson A et al (2002) Molecular biology of the cell. Taylor & Francis, New York
66. Loh KP, Bao Q et al (2010) Graphene oxide as a chemically tunable platform for optical applications. Nat Chem 2(12):1015

FSC
www.fsc.org
MIX
Papier aus verantwortungsvollen Quellen
Paper from responsible sources
FSC® C105338